T0214212

Communications
in Computer and Information Science   1264

More information about this series at http://www.springer.com/series/7899

Mostafa Belkasmi · Jalel Ben-Othman ·
Cheng Li · Mohamed Essaaidi (Eds.)

Advanced Communication Systems and Information Security

Second International Conference, ACOSIS 2019
Marrakesh, Morocco, November 20–22, 2019
Revised Selected Papers

 Springer

Editors
Mostafa Belkasmi 🆔
Mohammed V University
Rabat, Morocco

Jalel Ben-Othman 🆔
Centrale Supélec and Sorbonne University
Paris, France

Cheng Li 🆔
Memorial University of Newfoundland
St. John's, NL, Canada

Mohamed Essaaidi 🆔
Mohammed V University
Rabat, Morocco

ISSN 1865-0929   ISSN 1865-0937 (electronic)
Communications in Computer and Information Science
ISBN 978-3-030-61142-2   ISBN 978-3-030-61143-9 (eBook)
https://doi.org/10.1007/978-3-030-61143-9

This Springer imprint is published by the registered company Springer Nature Switzerland AG
The registered company address is: Gewerbestrasse 11, 6330 Cham, Switzerland

Preface

Welcome to the proceedings of the Second International Conference on Advanced Communication Systems and Information Security (ACOSIS 2019). The international conference was held during November 20–22, 2019, in the historic medieval city of Marrakesh, Morocco, one of the world's most important artistic and cultural centers. ACOSIS 2019 was co-organized by ENSIAS-Mohammed V University in Rabat, Morocco, FST-Hassan I University, Morocco, and the Association of Research and Innovation in Science and Technology, Morocco.

The purpose of ACOSIS 2019 was to provide a program for world research leaders and practitioners, to present and discuss their latest research results, ideas, developments, and applications in all areas of communication systems and information security.

ACOSIS 2019, by its second edition at a conference level, attracted a high volume of submitted papers covering various areas of communication systems, information security, and networking, coming from different countries. The conference received 94 paper submissions, out of which 20 papers were accepted, based on their readability, timelines and relevance, novelty and originality, technical content, and clarity of presentation, for inclusion in these proceedings.

Many people collaborated and worked hard to produce a successful ACOSIS 2019 conference. First of all, we thank all the authors for submitting their original contribution to the conference and for their presentations and discussions during the conference. Our sincere thanks go to the Program Committee members and the reviewers, who have done the hardest job of carefully reviewing the submitted papers. This event was an opportunity to invite international well-known personalities who accepted to present their last research. Our thanks go to our keynote and tutorial speakers: Behnaam Aazhang, Rice University, USA; Erol Gelenbe, IITIS-PAN and Imperial College London UK; Aria Nosratinia, The University of Texas at Dallas, USA; A. Murat Tekalp, Koc University, Turkey; Peter J. Tonellato, University of Missouri, USA; Jitendra K. Tugnait, Auburn University, USA; and Mohamed-Slim Alouini, KAUST, Saudi Arabia. Finally, we warmly thank all the partners, sponsors, and Organizing Committee members of ACOSIS 2019 for helping us formulate a rich technical program.

In the end, we truly hope that you will find the contributions presented in this volume interesting and stimulating. Enjoy!

July 2020

Mostafa Belkasmi
Jalel Ben-Othman
Cheng Li
Mohamed Essaaidi

Organization

Honorary Chair

Mohamed Slim Alouini   KAUST, Saudi Arabia

General Chairs

Mostafa Belkasmi   Mohammed V University in Rabat, Morocco
Mohamed Essaaidi   Mohammed V University in Rabat, Morocco

Executive Chairs

Idriss Chana   Moulay Ismail University, Morocco
Khalid Zine Dine   Mohammed V University in Rabat, Morocco

Keynotes Chairs

Mohamed Slim Alouini   KAUST, Saudi Arabia
Mustapha Benjilali   INPT Rabat, Morocco

Local Arrangement Chairs

Mustapha Elharoussi   Hassan I University, Morocco
Abdelmajid Hajami   Hassan I University, Morocco

Sponsorship Chairs

Hakim Allali   Hassan I University, Morocco
Ahlam Berkani   Mohammed V University in Rabat, Morocco

Session Chairs

Driss Benhaddou   University of Houston, USA
Hanan El Bakkali   Mohammed V University in Rabat, Morocco

Tutorial Chairs

Marwane Ayaida   University of Reims Champagne-Ardenne, France
Abdellatif Kobbane   Mohammed V University in Rabat, Morocco
Abdallah Rhattoy   Moulay Ismail University, Morocco

TPC Chairs

Jalel Ben-Othman	CentraleSupélec and Université Sorbonne Paris Nord, France
Cheng Li	Memorial University of Newfoundland, Canada

Track Chairs

Ahmed Azouaoui	Chouaib Doukkali University, Morocco
Abdelmalek Benzekri	Paul Sabatier University, France
Driss Bouzidi	Mohammed V University in Rabat, Morocco
Driss El Ouadghiri	Moulay Ismail University, Morocco
Alexander Gelbukh	Instituto Politécnico Nacional, Mexico
Shuai Han	Harbin Institute of Technology, China
Abdelkrim Haqiq	Hassan I University, Morocco
HyunBum Kim	University of North Carolina at Wilmington, USA
Pascal Lorenz	University of Upper Alsace, France
Lynda Mokdad	Paris-Est Créteil University, France
Sam Goundar	Victoria University of Wellington, New Zealand
Ahmad Taher Azar	Benha University, Egypt

TPC Members

Abdelkrim Abdelli	University of Science and Technology Houari Boumediene, Algeria
Hassan Abdelmounim	Hassan I University, Morocco
Sa'ed Abed	Kuwait University, Kuwait
Noreddine Abghour	Hassan II University, Morocco
Mohammed Chaouki Abounaima	Sidi Mohamed Ben Abdellah University, Morocco
Adnane Addaim	Ibn Tofail University, Morocco
Abdellah Adib	Hassan II University, Morocco
Yamine Ait Ameur	ENSEEIHT, France
Said Agoujil	Moulay Ismail University, Morocco
Dharma Agrawal	University of Cincinnati, USA
Majida Alasady	University of Tikrit, Iraq
Imane Allali	Mohammed V University in Rabat, Morocco
Najeeb Al-Shorbaji	eHealth Development Association, Jordan
El Miloud Ar-Reyouchi	Abdelmalek Essaadi University, Morocco
Meryeme Ayache	Mohammed V University in Rabat, Morocco
Bouchaib Aylaj	Chouaib Doukkali University, Morocco
Muhammad Reza Kahar Aziz	Institut Teknologi Sumatera, Indonesia
Mostafa Azizi	Mohammed I University, Oujda, Morocco
Mostafa Bellafkih	INPT Rabat, Morocco
Abdelhamid Belmekki	INPT Rabat, Morocco

Said Ben Alla	Hassan I University, Morocco
Mustapha Benjillali	INPT Rabat, Morocco
Ahlam Berkani	Mohammed V University in Rabat, Morocco
El Ouahidi Bouabid	Mohammed V University in Rabat, Morocco
Mohamed Boulouird	Cadi Ayyad University, Morocco
Abdellah Boulouz	Ibn Zohr University, Morocco
Jun Cai	Concordia University, Canada
Idriss Chana	Moulay Ismail University, Morocco
Mohamed Charkani	Sidi Mohamed Ben Absellah University, Morocco
Yawen Chen	University of Otago, New Zealand
Stefano Chessa	Università di Pisa, Italy
Domenico Ciuonzo	Network Measurement and Monitoring (NM2), Italy
Driss Driouchi	Mohammed I University, Morocco
Jamal El Abbadi	Mohammed V University in Rabat, Morocco
Hanan El Bakkali	Mohammed V University in Rabat, Morocco
Said El Kafhali	Hassan I University, Morocco
Abderrahman El Kafil	Moroccan Society for Telemedicine and eHealth, Morocco
Ahmed El Khadimi	INPT Rabat, Morocco
Khalid El Makkaoui	Mohammed I University, Morocco
Abdelkarim EL Mouatassim	Ibn Zohr University, Morocco
Mustapha Elharoussi	Hassan I University, Morocco
Khalifa Elmansouri	Ibn Zohr University, Morocco
Abdeslam En-Nouaary	INPT Rabat, Morocco
Mohammed Essaid Riffi	Chouaib Doukkali University, Morocco
Abderrazak Farchane	Sultan Moulay Slimane University, Morocco
Hacene Fouchal	University of Reims Champagne-Ardenne, France
Weihuang Fu	Google, USA
Antoine Gallais	Inria Lille, France
Bishnu Gautam	Wakkanai Hokusei Gakuen University, Japan
Alireza Ghasempour	ICT Faculty, USA
Kamal Ghoumid	Mohammed I University, Morocco
Zouhair Guennoun	Mohammed V University in Rabat, Morocco
Frederic Guilloud	IMT Atlantique, France
Yassine Hadjadj-Aoul	University of Rennes 1, France
Abdelmajid Hajami	Hassan I University, Morocco
Mohamed Hanini	Hassan I University, Morocco
Moulay Lahcen Hasnaoui	Université Moulay Ismail, Morocco
Elhassane Ibn-Elhaj	INPT Rabat, Morocco
Khalil Ibrahimi	Ibn Tofail University, Morocco
Said Jabbour	CRIL, CNRS, d'Artois University, France
Atman Jbari	Mohammed V University in Rabat, Morocco
Ali Kartit	Chouaib Doukkali University, Morocco
Omar Khadir	Hassan II University, Morocco
Baseem Khan	Hawassa University, Ethiopia
Lyes Khoukhi	University of Technology of Troyes, France

Donghyun Kim	Kennesaw State University, USA
Abdellatif Kobbane	Mohammed V University in Rabat, Morocco
Mohammed-Amine Koulali	Mohammed I University, Morocco
Cherkaoui Leghris	Hassan II University, Morocco
An Li	Nanchang University, China
He Li	Muroran Institute of Technology, Japan
Ruidong Li	National Institute of Information and Communications Technology (NICT), Japan
Hui Lin	University of Nevada, Reno, USA
Marco Listanti	Sapienza University of Rome, Italy
Yassine Maleh	Hassan I University, Morocco
Zoubir Mammeri	Paul Sabatier University, France
Mbarek Marwan	Chouaib Doukkali University, Morocco
Samir Mbarki	Ibn Tofail University, Morocco
Mokhtaria Mesri	University of Laghouat, Algeria
Abdellatif Mezrioui	INPT Rabat, Morocco
Vojislav Mišić	Ryerson University, Canada
Oussama Mjihil	Hassan I University, Morocco
Lei Mo	Inria, France
Negar Mohammadi-Koushki	University of Tehran, Iran
Mohamed Moughit	Hassan I University, Morocco
Najem Moussa	Chouaib Doukkali University, Morocco
Hammadi Nait-Charif	Bournemouth University, UK
Ahmed Nait-sidi-Moh	University of Picardy Jules Verne, France
Nidal Nasser	Alfaisal University, Saudi Arabia
Sarmistha Neogy	Jadavpur University, India
Abderrahmane Nitaj	Caen-Normandy University, France
Said Nouh	Hassan II University, Morocco
Niyazi Odabasioglu	Istanbul University, Turkey
Ayoub Otmani	University of Rouen, France
Paulo Pires	Federal University of Rio de Janeiro, Brazil
Said Rakrak	Cadi Ayyad University, Morocco
Salma Rattal	Hassan II University, Morocco
Eric Renault	CNRS, ESIEE Paris, France
Muhammad Reza Kahar Aziz	Institut Teknologi Sumatera, Indonesia
Yassine Rhazali	Moulay Ismail University, Morocco
Mohammed Ridouani	Hassan II University, Morocco
Rahal Romadi	Mohammed V University in Rabat, Morocco
Abdel-Badeeh Salem	Ain Shams University, Egypt
Mohamed Sall	Université Alioune Diop de Bambey, Senegal
Fortunato Santucci	University of L'Aquila, Italy
Anna Schmausklughammer	University of Deggendorf, Germany
Bo Sheng	University of Massachusetts Boston, USA
El Mamoun Souidi	Mohammed V University in Rabat, Morocco

Omid Taghizadeh	RWTH Aachen University, Germany
Maen Takruri	American University of Ras Al Khaimah, UAE
Adil Tannouche	Sultan Moulay Slimane University, Morocco
Bulent Tavli	University of Economics and Technology, Turkey
A. Murat Tekalp	Koc University, Turkey
Ming-Fong Tsai	National United University, Taiwan
Hicham Tribak	Abdelmalek Essaadi University of Tetouan, Morocco
Hao Wang	Stanford University, USA
Ahmad Samer Wazan	IRIT Toulouse, France
Myounggyu Won	The University of Memphis, USA
Xuanli Wu	Harbin Institute of Technology, China
Yang Xiao	The University of Alabama, USA
Huiyue Yi	Shanghai Research Center for Wireless Communications, China
Yang Yi	Virginia Tech, USA
Mohamed Younas	Oxford Brookes University, UK
Wei Yuan	Huazhong University of Science and Technology, China
Abdellah Zaaloul	Ibn Zohr University, Morocco
Sherali Zeadally	University of Kentucky, USA
Nadia Zeghib	Constantine 2 University, Algeria
Ahmed Zellou	Mohammed V University in Rabat, Morocco
El Moukhtar Zemmouri	Moulay Ismail University, Morocco
Chaofeng Zhang	Muroran Institute of Technology, Japan
Haibo Zhang	Universtiy of Otago, New Zealand
Ping Zhou	Qualcomm, USA
Khalid Zine Dine	Mohammed V University in Rabat, Morocco

External Reviewers

Amine Abouaomar	Mohammed V University in Rabat, Morocco, and Université de Sherbrooke, Canada
Rajendran Sobha Ajin	Idukki District Emergency Operations Center, India
Mohammad Al Hattab	Al Ain University, UAE
Hakem Alazmi	Kennesaw State University, USA
Mariame Amine	Mohammed V University in Rabat, Morocco
Ali Azougaghe	Mohammed V University in Rabat, Morocco
Reda Benkhouya	Ibn Tofail University, Morocco
Abdelaali Chaoub	INPT Rabat, Morocco
Meryem Cherkaoui Semmouni	Mohammed V University in Rabat, Morocco
Jiong Dong	Muroran Institute of Technology, Japan
Otmane El Mouaatamid	Mohammed V University in Rabat, Morocco
Zakaria El-Moutaouakkil	Telecom Bretagne and Lab-Sticc, France
Hicham Grari	Chouaib Doukkali University, Morocco
Abdelali Hadir	Hassan ll University, Morocco

Hamza Boualame	Mohammed V University in Rabat, Morocco
Dimitra Ignatiadis	Stanford University, USA
Meriem Joundi	Hassan ll University, Morocco
Hicham Laanaya	Mohammed V University in Rabat, Morocco
Zouheir Labbi	Mohammed V University in Rabat, Morocco
Souad Labghough	Mohammed V University in Rabat, Morocco
Ahmed Maarof	Mohammed V University in Rabat, Morocco
Lahcen Niharmine	Mohammed V University in Rabat, Morocco
Noria Tahiri	Mohammed V University in Rabat, Morocco
Tianzhu Pan	Harbin Institute of Technology, China
Mohammed Raiss El Fenni	INPT Rabat, Morocco
Sujan Ray	University of Cincinnati, USA
Karim Rkizat	Mohammed V University in Rabat, Morocco
Hayat Sedrati	National School of Public Health, Morocco
Abhishek Shivanna	University of Cincinnati, USA
Wei Sun	University of Nebraska-Lincoln, USA
Walter Tiberti	University of L'Aquila, Italy
Haitham Yaish	Ajman University, UAE
Anouar Yatribi	Mohammed V University in Rabat, Morocco
Chengcheng Zhao	Shanghai Jiao Tong University, China

Contents

Wireless Communications and Services

Pilot Assignment vs Soft Pilot Reuse to Surpass the Pilot Contamination
Problem: A Comparative Study in the Uplink Phase 3
    Abdelfettah Belhabib, Mohamed Boulouird, and Moha M'Rabet Hassani

New Channel Estimation for MC-CDMA System Under Fast
Multipath Channel. 14
    Hocine Merah, Larbi Talbi, Mokhtaria Mesri, and Khaled Tahkoubit

Multi-party WebRTC as a Managed Service over Multi-Operator SDN 25
    Riza Arda Kirmizioglu and A. Murat Tekalp

NCBP: Network Coding Based Protocol for Recovering Lost Packets
in the Internet of Things 38
    El Miloud Ar-Reyouchi, Yousra Lamrani, Imane Benchaib,
    Salma Rattal, and Kamal Ghoumid

Vehicular Communications

Vehicle Collision Avoidance 53
    Khawla A. Alnajjar, Noora Abdulrahman, Fatma Mahdi,
    and Marwah Alramsi

On the V2X Velocity Synchronization at Unsignalized Intersections:
Right Hand Priority Based System 65
    Saif Islam Bouderba and Najem Moussa

Convolutional Neural Networks for Traffic Signs Recognition 73
    Btissam Bousarhane and Driss Bouzidi

Channel Coding

An Efficient Semi-Algebraic Decoding Algorithm for Golay Code 95
    Hamza Boualame, Idriss Chana, and Mostafa Belkasmi

Iterative Decoding of GSCB Codes Based on RS Codes Using
Adapted Scaling Factors. 103
    Es-said Azougaghe, Abderrazak Farchane, Said Safi,
    and Mostafa Belkasmi

Turbo Decoder Based on DSC Codes for Multiple-Antenna Systems   115
Abdelghani Boudaoud, Mustapha El Haroussi,
and Elhassane Abdelmounim

A Threshold Decoding Algorithm for Non-binary OSMLD Codes   121
Zakaria M'rabet, Fouad Ayoub, and Mostafa Belkasmi

Construction of Error Correcting Codes

On the Hamming and Symbol-Pair Distance of Constacyclic Codes
of Length p^s over $\mathbb{F}_{p^m} + u\mathbb{F}_{p^m}$ 137
Jamal Laaouine

Intrusion Detection Techniques

A Comparison of Different Machine Learning Algorithms
for Intrusion Detection   157
Basma Karbal and Rahal Romadi

A Weighted LSTM Deep Learning for Intrusion Detection   170
Meryem Amar and Bouabid EL Ouahidi

Proposed Solution for HID Fileless Ransomware Using Machine Learning . . .   180
Mohamed Amine Kerrich, Adnane Addaim, and Loubna Damej

Wireless and Mobile Network Security

Data Oriented Blockchain: Off-Chain Storage with Data Dedicated
and Prunable Transactions   195
Oualid Boumaouche, Afifa Ghenai, and Nadia Zeghib

A Framework to Secure Cluster-Header Decision in Wireless Sensor
Network Using Blockchain   205
Hafsa Benaddi, Khalil Ibrahimi, Haytham Dahri,
and Abderrahim Benslimane

Security in MANETs: The Blockchain Issue   219
Nada Mouchfiq, Chaimae Benjbara, and Ahmed Habbani

Improving IoT Security with Software Defined Networking (SDN)   233
Abdelaali Tioutiou and Ouafaa Diouri

Applied Cryptography

Optical Image Encryption Process Using Triple Deterministic Spherical
Phase Masks Array   241
   Wiam Zamrani and Esmail Ahouzi

Author Index   251

Wireless Communications and Services

Pilot Assignment vs Soft Pilot Reuse to Surpass the Pilot Contamination Problem: A Comparative Study in the Uplink Phase

Abdelfettah Belhabib[1], Mohamed Boulouird[1,2]($\boxtimes$),
and Moha M'Rabet Hassani[1]

[1] Instrumentation, Signals and Physical Systems (I2SP) Group,
Faculty of Sciences Semlalia, Cadi Ayyad University, Marrakesh, Morocco
`abdelfettah.belhabib@edu.uca.ac.ma`, `m.boulouird@uca.ac.ma`,
`hassani@ucam.ac.ma`
[2] National School of Applied Sciences of Marrakesh (ENSA-M),
Cadi Ayyad University, Marrakesh, Morocco

Abstract. Even though, Massive MIMO (M-MIMO) technology offers good improvement in terms of performance and quality of communications between users and Base Stations (BS). This technology, still limited by a harmful constraint renowned Pilot Contamination (PC) problem. To beat this problem of PC, two approaches have been adapted as a medicine for this problem, which appears strongly in Multi-Cell M-MIMO systems. The first approach is based on assigning extra pilot sequences to different users, while the second approach obliges the reuse of the same set of pilots in different cells. This paper reviews, briefly, the problem of PC in M-MIMO systems. Thereafter, it analyzes and provides a comparison between two decontaminating strategies: the Soft Pilot Reuse (SPR) and the Pilot Assignment based on the Weighted Graph Coloring (WGC-PA), which are respectively based on the above-discussed approaches. The analysis presented in this paper is focused on the uplink phase (i.e reverse link).

Keywords: Massive MIMO · Pilot Contamination · Soft Pilot Reuse · Weighted Graph Coloring · Pilot assignment · 5G wireless communications

1 Introduction

Based on the studies of CISCO [1], data traffic tends to reach a very high peak between 2017 to 2022, where mobile-connected tablets and phones will generate 6.8 GB of traffic per month on average compared to 2017 where it was just 3.3 GB per month on average. These remarkable exceeds on term of data traffics, makes on 5G the only available choice to control the manipulation of this increases

© Springer Nature Switzerland AG 2020
M. Belkasmi et al. (Eds.): ACOSIS 2019, CCIS 1264, pp. 3–13, 2020.
https://doi.org/10.1007/978-3-030-61143-9_1

of data traffic, [2] proposes a very interesting technology to achieve the fifth generation, this technology is called M-MIMO, which is mainly based on the use of hundreds of antennas at BS to serve simultaneously many users (i.e dozens of users) [3]. However, as the available set of orthogonal pilot sequences (i.e band of frequencies within a specific time slots) is limited, compared to the number of users in the overall cells, it is necessary to reuse the same set of pilots to serve the users of all cells, which lead to the problem of interference, which degrade the quality-of-service offered to users [4]. The interferences either inter-cell interferences or intra-cell interferences or both of them are generated between users that reuse the same pilots; these overlaps refer to the phenomenon known as PC [5,6], which is considered as the main bottleneck that limits the fulfillment of the expected benefits of M-MIMO technology [2]. The PC problem is depicted in Fig. 1, where two users of two different cells corrupt each other because they are set to reuse the same pilot sequence.

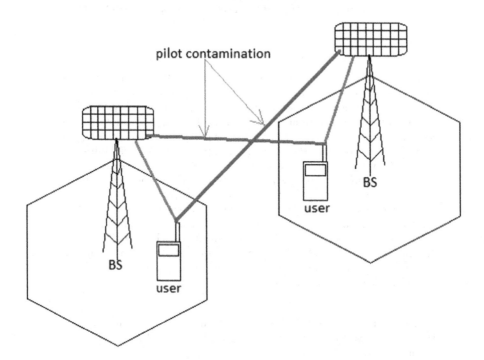

Fig. 1. The problem of pilot contamination

Regarding the fact that the quality of communication between users and their supported BSs, depends on the accuracy of the estimated channel state information (CSI) within the training phase [2]. Therefore, it is easy to note that the PC leads to an inaccurate (CSI), moreover as the detectors and precoders (used respectively for data detection and data precoding) are both, built based on the estimated channels (i.e CSIs), hence, the obtained achievable rates of users are low in both the uplink (UL) and downlink (DL) phases, moreover the achievable rates saturate whatever the number of antennas deployed in BS [7].

To address the problem of PC within M-MIMO systems, several approaches have been proposed to cancel/reduce this problem of PC [4,8]. Among these approaches we cite: the approach which is based on assigning extra orthogonal pilot sequences (e.g. SPR) and the approach which uses fewer pilots (e.g. WGC-PA). Both of these approaches are powerful, but none of them take into account the problem of pilot overhead and they arbitrary use a specific set of orthogonal pilots. Hence, the contributions of this paper are:

- Review the problem of PC in M-MIMO systems.
- Show the degradable effect of the problem of pilot overhead on the effectiveness of some decontaminating strategies. That, through studying and comparing two decontaminating strategies SPR and WGC-PA which are based on the two above discussed approaches.

The present paper is organized as follows. The related work is presented in Sect. 2. Section 3 presents the system model for M-MIMO technology which will be adopted within this paper. Section 4 provides an overview of the problem of pilot contamination. Section 5 analyzes and provides a comparison between the SPR and the WGC-PA strategies on the uplink phase. In Sect. 6, simulation results provide a comparison between the performances of the conventional strategy (CS), the SPR, and the WGC-PA strategies. Section 7 finally, gives a conclusion.

2 Related Work

During the past decade, researchers suggest some solutions to beat the problem of PC, in [9] the coherence time interval is divided into two slots, which are used separately for both the channel training and the data transmission phases, regarding the impact of the number of the employed pilots on the obtained achievable rates, [9] manipulates two metrics to maximize the achievable rate, which are: the length of pilot sequences, and the pilot reuse factor (i.e sharing the available pilots between cells in what it called by pilot assignment strategy) that by assuming an infinite number of antennas BSs, therefore [10,11] aims to reduce the problem of PC through increasing the distance between cells that reuse the same set of pilots, specifically cells are clustered into groups based on optimizing the pilot reuse factor, thereafter pilots are optimally assigned to the cells of each cluster regarding the goal of maximizing the achievable rates, on the other side [11] focuses on separating the users within cells into different groups based on their need to manipulate either high or low data traffic, thereafter [11] derives a closed-form solution by optimizing the achievable rates for a given number of pilot sequences; [9,11] commonly assume having a number of pilots larger than the number of users per cells, which require extra pilot sequences, moreover the users belonging to the same cell are not permitted to reuse the same pilot sequences (i.e null intra-cell interference); [12] aims to mit-igate the PC through a pilot assignment strategy, where users of edge cells are considered having a low signal-to-interference-plus-noise ratio (SINR) and must be assigned by orthogonal pilots to reduce the strength of inter-cell interferences, while center users with high SINRs are obliged to reuse the same set of pilots within different cells, specifically [12] arranges the users within cells based on their SINR -at the users side- into either edges or centers, thereafter a threshold of distance for classification is derived by approximating the SINR as a function of distance, then the optimal distance is obtained through the optimization of the network throughput, it should be noted that the SINR is derived based on an infinite number of antennas BSs, in addition the noise is omitted from the SINR expressions, moreover the derived SINR is based on a circular shape of cell, which is not the case in practice and the problem of pilot overhead had not taken in account when the achievable rate is derived. [13] proposes a coordinated approach between users and their corresponding BSs to accurately estimate the CSIs within each BS, regarding the fact that the number of BSs is too less com-pared to the number of users; therefore, BSs are assigned by orthogonal pilots, which are broadcasting in the downlink phase, therefore users in each cell exploit these broadcasted pilots to estimate the downlink channels, thereafter users of each cell synchronously uplink the same set of pilots (i.e full pilot reuse factor) then BSs estimate the channels based on the received pilots, thereafter users again uplink their pilots accompanied with the estimated channels, BSs exploit the received signals to secondly estimate the CSIs, therefore BSs remove the first contaminated signal from the second one to obtain the accurate CSIs; however this decontaminating strategy is not desired for such practical system because it requires much time for training which may degrade the latency of the wireless communication system, moreover it assumes a non-overlapping angle-of-arrivals (AoAs).

3   Massive MIMO System Model

The system model adopted within the present paper is composed of L hexagonal cells, where each cell contains K single antenna users served by a BS having M antennas. Let $g_{m,j,k,l}$ be the propagation factor of the kth user in the lth cell to the mth antenna of the BS belonging to the jth cell, then $g_{m,j,k,l}$ can be split into the product of two terms as:

$$g_{m,j,k,l} = h_{m,j,k,l} \sqrt{\beta_{j,k,l}} \tag{1}$$

where $\beta_{j,k,l}$ denotes the slow-scale-fading (SSF) factor and it accounts for the path loss, and the shadow fading, while $h_{m,j,k,l}$ denotes the fast-fading coef-ficient (FF) which is considered to be i.i.d (i.e independent and identically distributed) having a Gaussian distribution of zero-mean and unity variance $\sim \mathcal{CN}(0,1)$. Notice that, for the reason of simplicity, the antennas index m in (1) is dropped away. Therefore the SSF of a user k in the lth cell to the BS of the jth cell can be re-written as in [14] as:

$$\beta_{j,k,l} = \frac{z_{j,k,l}}{(r_{j,k,l}/R)^\gamma} \qquad (2)$$

where $r_{j,k,l}$ is the distance between the kth user of the lth cell to the BS of the jth cell, γ denotes the signal decay exponent (i.e path loss exponent), and $z_{j,k,l}$ is a log-normal random distribution. Therefore, the channel matrix of the K users of the lth cell to the BS of the jth cell can be expressed as in [14] as follows:

$$G_{j,l} = [g_{j,1,l} g_{j,2,l}, ..., g_{j,K,l}] = [h_{j,1,l}, h_{j,2,l}, ..., h_{j,K,l}] D_{j,l}^{\frac{1}{2}} \qquad (3)$$

where $D_{j,l}$ is a diagonal matrix, and it can be expressed as :

$$D_{j,l} = \begin{pmatrix} \beta_{j,1,l} & 0 & ... & 0 \\ & & & \cdot \\ 0 & \beta_{j,2,l} & & \cdot \\ & & \cdot & \\ 0 & ... & ... & \beta_{j,K,l} \end{pmatrix} \qquad (4)$$

4   Pilot Contamination

The fact that the coherence interval (i.e time-frequency domain) is short and limited compared to the total number of the active users within cells, obliges the reuse of the same set of orthogonal pilot sequences during the channel training phase, to serve different users of the adjacent cells. Consequently, the interferences are generated between the users that employ the same pilots, leading to the problem of PC; therefore, to beat this constraint, $K.L$ orthogonal pilots are required, which is not possible in practice because the coherence time interval and the system bandwidth are both limited [7,13]. To completely avoid the problem of contamination during the training phase, the orthogonality condition must be satisfied, which can be expressed as in [15] as:

$$\phi_{k,i}^H \phi_{k',j} = \delta[k - k']\delta[i - j] \qquad (5)$$

Where

$$\delta[k - k']\delta[i - j] = \begin{cases} 1 & \text{if } k = k' \text{ and } i = j \\ 0 & \text{otherwise} \end{cases} \qquad (6)$$

Where $(.)^H$ refers to the transpose conjugate of a matrix, $\phi_{k,i} : \{(i,j) = (1, 2, ..., L), k = 1, 2, ..., K\}$ is the pilot sequence of length τ of the kth user in the ith cell, which is a $\tau \times 1$ vector. (5) must be satisfied to avoid the problem of PC. However, the scarcity of the available pilots and the limited coherence interval impose the use of non-orthogonal pilot sequences, which lead to the problem of PC.

5   SPR and WGC-PA Strategies

5.1   SPR Strategy

Owing to the fact that, users within cells should not be treated as they are similar, [14] asserts that the users which are close to their corresponding BSs, do not suffer from a high effect of PC as the outer users, that why the SPR strategy suggests to divide the users within cells -based on their SSF- into two sets, which are labeled outer and inner users. The K users of the jth cell, are then divided into two groups as in [14] based on the following threshold:

$$\rho_j = \frac{\lambda}{K_{CS}} \sum_{k=1}^{K} \beta_{j,k,j}^2 \tag{7}$$

where $K_{CS} = max(k_j) : j = 1, 2, ..., L$ is the number of the required pilots in the CS [2], λ is a parameter used to control the system configuration. Therefore, users are considered either inner or outer based on the condition:

$$\beta_{j,k,j}^2 > \rho_j \rightarrow \begin{cases} Yes & \text{inner user,} \\ No & \text{outer user} \end{cases} \tag{8}$$

Therefore, based on both (7) and (8), users are divided and considered to be either, outer or inner users. Then, outer users of all cells are assigned by orthogonal pilots while inner users are obliged to reuse the same set of pilot sequences within different cells. Accordingly, the number of pilot sequences required in the SPR strategy is:

$$K_{SPR} = \sum_{j=1}^{L} K_j^e + \max\{K_j^c : j = 1, 2, ..., L\} \tag{9}$$

Here, K_j^e, K_j^c denote respectively, the number of outer and inner users in the jth cell. Therefore, based on the least square estimator (LS), the estimated channels by the BS of the jth cell is given as in [14] as:

$$\begin{cases} \hat{G}_{j,j}^c = G_{j,j}^c + \frac{N_j^p}{\sqrt{\rho_p}} \phi_c^H \\ \hat{G}_{j,j}^e = G_{j,j}^e + \sum_{l \neq j}^{L} G_{j,l}^e \phi_{e,l} \phi_{e,j}^H + \frac{N_j^p}{\sqrt{\rho_p}} \phi_{e,j}^H \end{cases} \tag{10}$$

where ϕ_c, $\phi_{e,j}$ denote, the pilot sequences of length τ, which are used respectively for the inner users and for the outer users of the jth cell, N_j^p is the additive white Gaussian noise (AWGN) upon the antennas BS of the jth cell. Therefore, the SINR of the kth user in the jth cell is:

$$SINR_{j,k}^{u,e} = \frac{\rho_u |(g_{j,k,j}^e)^H g_{j,k,j}^e|^2}{\rho_u \sum_{k' \neq k} |(g_{j,k,j}^e)^H g_{j,k',j}^e|^2 + |inter_{j,k}^{u,e}|^2} \tag{11}$$

$$SINR_{j,k}^{u,c} = \frac{\rho_u |(g_{j,k,j}^c)^H g_{j,k,j}^c|^2}{\rho_u \sum_{l \neq j} |(g_{j,k,l}^c)^H g_{j,k,l}^c|^2 + |inter_{j,k}^{u,c}|^2} \tag{12}$$

where $inter_{j,k}^{u,e}$, $inter_{j,k}^{u,c}$ are the interference components upon the kth user of the jth cell, regarding the group where it belong to (i.e outer or inner user). Then, the achievable rate on the uplink phase is expressed as:

$$\begin{cases} C_{j,k}^{u,e} = (1 - \frac{K_{SPR}}{K_{CS}}\mu)E\{log_2(1 + SINR_{j,k}^{u,e})\} \\ C_{j,k}^{u,c} = (1 - \frac{K_{SPR}}{K_{CS}}\mu)E\{log_2(1 + SINR_{j,k}^{u,c})\} \end{cases} \tag{13}$$

where $0 < \mu < 1$ denotes a parameter that evaluates the loss of the available bandwidth, which is caused by using pilots in the channel estimation phase.

5.2   WGC-PA

Instead of the SPR strategy, the WGC-PA treats the users within cells as they have similar characteristics. Accordingly, it constructs a graph known as the interference graph, where the users are considered to be the vertexes of this graph, and for each pair of users, the strength of interference is computed based on a specific parameter χ. For example, the strength of interference generated between two users k and k', which belong respectively to the jth cell and the lth cell is expressed as:

$$\chi_{(j,k),(l,k')} = \frac{\beta_{j,k',l}^2}{\beta_{j,k,j}^2} + \frac{\beta_{l,k,j}^2}{\beta_{l,k',l}^2} \tag{14}$$

Then, the users are assigned with different orthogonal pilot sequences to minimize the interference parameter (14). In other words, minimizing the parameter $\chi_{(j,k),(l,k')}$ to achieve a maximum achievable rate. The derived SINR of the kth user in the jth cell is given in [16] as follows:

$$SINR_{(j,k)}^u = \frac{\rho_u |(g_{(j,k),j})^H g_{(j,k),j}|^2}{\rho_u \sum_{k' \neq k} |(g_{(j,k),j})^H g_{(j,k'),j}|^2 + |inter_{(j,k)}^u|^2} \tag{15}$$

where $inter_{(j,k)}^u$ denotes the interference component upon the considered user k of cell j. Then; the main goal of the WGC-PA strategy is to reduce the strength of PC upon users, which reuse the same pilots, that by allocating different pilots to users which are supposed to suffer on a higher level of PC. Therefore; the derived achievable rate of the kth user in the jth cell, can be expressed as in [16] as:

$$C_{(j,k)}^u = (1 - \frac{S}{K_{CS}}\mu)E\{log_2(1 + SINR_{(j,k)}^u)\} \tag{16}$$

where S is the number of orthogonal pilot sequences used by the WGC-PA strategy.

5.3   Comparison Between the Two Strategies

From the analysis above, it is recognized that the SPR strategy treats the users of each cell separately. Hence, all outer users are assigned with different orthogonal pilots. In other words, each outer user is assigned with its specific pilot. However; as the number of outer users increases, the requirement of orthogonal pilot sequences increases too, which can lead to the problem of pilot overhead. In fact, the bandwidth left for data transmission becomes tight. Therefore, degrade the achievable rates. On the other side, the WGC-PA strategy aims to reduce the impact of contamination upon users that reuse the same pilot sequences. But as the number of users within cells increases, the WGC-PA strategy tends to perform as the CS, where users are randomly assigned by the same set of orthogonal pilot sequences. Even though; SPR and WGC-PA are used to mitigate the problem of PC within M-MIMO systems, they both become weaker when the number of users per cell, is larger than the number of the available pilot resources. Moreover, SPR spents large amounts of pilots.

6   Simulation Results

To evaluate the performances of both, the SPR and the WGC-PA strategies, a set of Monte Carlo simulation of 100 iterations is considered, which is applied to the Multi Cellular M-MIMO system described in the Sect. 3, and the parameters are fixed as in Table 1.

Table 1. System settings

Number of cells	$L = 7$
Number of antennas BS	$32 \leq M \leq 256$
Number of users per cell	$K = 12$
Adjustment parameter λ	$0.05 \leq \lambda \leq 1$
Cell radius	$R = 500\,\mathrm{m}$
Inner radius	$r = 30\,\mathrm{m}$
Log normal shadow fading	$\sigma_{shadow} = 8\,\mathrm{dB}$
Transmit power in UL & DL ρ_p, ρ_u and ρ_d	$15\,\mathrm{dBm}$
Path loss exponent	$\gamma = 3$
Pilot overhead parameter	$0 < \mu < 1$

Figure 2 shows the average of the uplink achievable rate (e.g. $C_j = \sum_{k=1}^{K} C_{j,k}^u$ for the case (16)) against the number of antennas BSs, based on two low complexity detectors, which are the zero-forcing (ZF) and the Matched filter (MF) detectors. The performances of WGC-PA, slightly exceed those of the CS in both detectors, that because the number of users per cell is large (i.e $K = 12$ user per cell and $K \times L = 84$ user per seven cells), and the WGC-PA strategy tends to perform as the CS for a large number of users. On the other side, the SPR strategy performs better than the WGC-PA by about 9 bps/Hz and 2 bps/Hz respectively for ZF and MF detectors, however, as the number of antennas BSs increases, the SPR strategy tends to lose its brilliance compared to the WGC-PA strategy.

Figure 3 shows the effect of allowing more orthogonal pilots to the WGC-PA strategy, compared to the results depicted in Fig. 2. The most important remarque is that the performance of the WGC-PA strategy is improved by about 5 bps/Hz and 1 bps/Hz respectively in the case of ZF and MF detectors. Accordingly, WGC-PA can exceeds SPR when it is allowed to exploits large amounts of pilots as the SPR strategy.

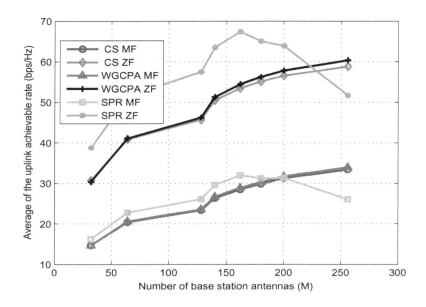

Fig. 2. The average achievable rate on the uplink phase, $\lambda = 0.05, S = K_{CS}$

6.1 Further Discussion

From Fig. 2 and Fig. 3, it is well seen that the number of pilots used by the two decontaminating techniques affects their performances, thus, the SPR strategy requires the use of different pilots to all outer users, which may leads to narrowing the bandwidth left for data transmission, while WGC-PA try to exploit a specific set of pilots to reduce the impact of PC. Therefore, the performances of WGC-PA tends to exceed the SPR if it is allowed to use additional pilot-resources. Consequently, the choice of such a decontaminating technique must depend on some parameters as the available pilot resources and the number of users per cell.

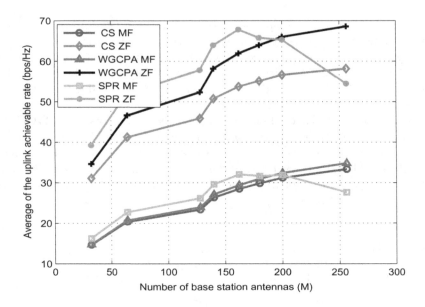

Fig. 3. The average achievable rate on the uplink phase, $\lambda = 0.05, S = 25$

7 Conclusion and Future Works

During the present paper, the problem of pilot contamination within the Multi-cell M-MIMO system is briefly reviewed. Thereafter, two decontaminating strategies are analyzed and compared to each other on the uplink phase to show the impact of pilot overhead on such decontaminating technique. These strategies are the soft pilot reuse (SPR) and the weighted graph coloring pilot assignment (WGC-PA), the main idea behind choosing these strategies is the fact that they are based on two different approaches (i.e the approach that requires the use of additional pilots during the training phase and the other approach which is based on the use of a specific set of pilots). Simulation results show that the number of pilot sequences used in the training phase affects the performances obtained by both the SPR and the WGC-PA strategy, and the WGC-PA perform better than the SPR strategy when it is allowed to employing a set of pilots larger than the number of users per cell. Hence, it is necessary to take into account the problem of pilot overhead before providing such a decontaminating technique.

References

1. Cisco visual networking index: global mobile data traffic forecast update, 2017–2022. White paper, Cisco public, February 2019. https://s3.amazonaws.com/media.mediapost.com/uploads/CiscoForecast.pdf
2. Marzetta, T.L.: Noncooperative cellular wireless with unlimited numbers of base station antennas. IEEE Trans. Wirel. Commun. **9**(11), 3590–3600 (2010)

3. Riadi, A., Boulouird, M., Hassani, M.M.: An overview of massive-MIMO in 5G wireless communications. In: Colloque International TELECOM 2017 & 10èmes JFMMA, EMI - Rabat, Morocco, Mai, pp. 10–12 (2017)
4. Belhabib, A., Boulouird, M., Hassani, M.M.: Mitigation techniques of pilot contamination in massive MIMO systems for 5G wireless communications: an overview. In: Colloque International TELECOM 2019 & 11èmes JFMMA, Saidia, Morocco, 12–14 June 2019
5. Wang, M., Wang, D.: Sum-rate of multi-user MIMO systems with multi-cell pilot contamination in correlated Rayleigh fading channel. Entropy **21**(6), 573 (2019)
6. Al-hubaishi, A., Noordin, N.K., Sali, A., Subramaniam, S., Mansoor, A.M.: An efficient pilot assignment scheme for addressing pilot contamination in multicell massive MIMO systems. Electronics **8**(4), 372 (2019)
7. Elijah, O., Leow, C.Y., Rahman, T.A., Nunoo, S., Iliya, S.Z.: A comprehensive survey of pilot contamination in massive MIMO-5G system. IEEE Commun. Surv. Tutor. **18**(2), 905–923 (2016). Secondquarter
8. Belhabib, A., Boulouird, M., Hassani, M.M.: The impact of using additional pilots on the performance of massive MIMO systems. In: 2019 International Conference on Signal, Control and Communication (SCC), Hammamet, Tunisia, pp. 87–92 (2019). https://doi.org/10.1109/SCC47175.2019.9116177
9. Sohn, J.Y., Yoon, S.W., Moon, J.: When pilots should not be reused across interfering cells in massive MIMO. In: 2015 IEEE International Conference on Communication Workshop (ICCW), London, pp. 1257–1263 (2015)
10. Sohn, J., Yoon, S.W., Moon, J.: On reusing pilots among interfering cells in massive MIMO. IEEE Trans. Wirel. Commun. **16**(12), 8092–8104 (2017)
11. Sohn, J., Yoon, S.W., Moon, J.: Pilot reuse strategy maximizing the weighted-sum-rate in massive MIMO systems. IEEE J. Sel. Areas Commun. **35**(8), 1728–1740 (2017)
12. Fan, J., Li, W., Zhang, Y.: Pilot contamination mitigation by fractional pilot reuse with threshold optimization in massive MIMO systems. Digit. Sign. Process. **78**, 197–204 (2018)
13. Zhang, J., Zhang, B., Chen, S., Mu, X., El-Hajjar, M., Hanzo, L.: Pilot contamination elimination for large-scale multiple-antenna aided OFDM systems. IEEE J. Sel. Top. Sign. Process. **8**(5), 759–772 (2014)
14. Zhu, X., et al.: Soft pilot reuse and multicell block diagonalization precoding for massive MIMO systems. IEEE Trans. Veh. Technol. **65**(5), 3285–3298 (2016)
15. Boulouird, M., Riadi, A., Hassani, M.M.: Pilot contamination in multi-cell massive-MIMO systems in 5G wireless communications. In: 2017 International Conference on Electrical and Information Technologies (ICEIT), pp. 1–4 (2017). https://doi.org/10.1109/EITech.2017.8255299
16. Zhu, X., Dai, L., Wang, Z., Wang, X.: Weighted-graph-coloring-based pilot decontamination for multicell massive MIMO systems. IEEE Trans. Veh. Technol. **66**(3), 2829–2834 (2017)

New Channel Estimation for MC-CDMA System Under Fast Multipath Channel

Hocine Merah[1], Larbi Talbi[2(✉)], Mokhtaria Mesri[1], and Khaled Tahkoubit[3]

[1] Laboratoire des Semi Conducteurs et Matériaux Fonctionnels,
Laghouat University, 03000 Laghouaty, Algeria
merah.hossein2@gmail.com, meradmesri@yahoo.fr
[2] Department of Computer Science and Engineering, Université du Québec-UQO,
Gatineau, QC, Canada
larbi.talbi@uqo.ca
[3] Electronics Department, University of Sciences and Technology of Oran,
El Mnaouar, BP 1505, Bir El Djir, 31000 Oran, Algeria
khaled.tahkoubit@univ-usto.dz

Abstract. One of the basic keys in the next generation wireless communications systems is to propose a convenient channel estimation technique, which in turn provides a better Bit Error Rate (BER). In Multicarrier Code Division Multiple Access (MC-CDMA) system, channel estimation is very important to overcome the effect of channel fading, which may distort the symbols of the information and resulting in a considerable degradation in the BER. On the other hand, the digital modulator and demodulator associated with MC-CDMA modulation is an acceptable numerical complexity (logarithmic) for applications with low computing resources. In this paper, a New Channel Estimation (NCE) scheme has been proposed by reserving suitable positions for the pilot carrier, mainly based on interleaved partitions, by filling these carriers with the unit signal to avoid the computational complexity involved in the channel estimation scheme. The simulation results show that the NCE method applied to MC-CDMA system achieves a better improvement regarding BER performance under Rayleigh and Rice fading channels.

Keywords: BER · MC-CDMA · NCE · Fading channels · Rayleigh · Rice

1 Introduction

Multi-carrier Code Division Multiple Access (MC-CDMA) [2] is based on the combination of multi carrier modulations and spread spectrum using Code Division Multiple Access (CDMA). Since its appearance in 1993 [3], MC-CDMA has been the subject of numerous comparisons with systems using the Direct-Sequence Code Division Multiple Access (DS-CDMA) technique [10]. These comparisons have largely demonstrated the superiority of MC-CDMA systems towards DS-CDMA systems. In addition to these comparative studies, works,

© Springer Nature Switzerland AG 2020
M. Belkasmi et al. (Eds.): ACOSIS 2019, CCIS 1264, pp. 14–24, 2020.
https://doi.org/10.1007/978-3-030-61143-9_2

mainly carried out on a downlink, have sought to optimize MC-CDMA systems in order to improve their performance. Thus, the sensitivities of the MC-CDMA systems with respect to Doppler shifts or synchronization errors were assessed. New detection techniques have emerged and compared to commonly used techniques. When the Walsh-Hadamard codes [9] are chosen, it is possible to perform the spreading function and the Fourier transform in a single operation, thus reducing the complexity of the MC-CDMA transmitters.

The classifications of MC-CDMA receptors are carried out according to different criteria which relate to the structure of the receiver (series, parallel, feedback and so on), as well as the equalization technique that is used. Consequently, optimal receivers can be differentiated from sub optimal receivers, linear receivers from non-linear receivers, multi-user receivers from single-user receivers, and so on [1]. In the receiver of a multi-carrier system such as the MC-CDMA system, single-user or multi-user detection techniques are implemented to detect the wanted signal. These techniques, also called equalization techniques in the particular case of MC-CDMA systems, are intended to compensate for the amplitude and phase distortions introduced by the transmission channel. However, to ensure a satisfactory transmission quality, these distortions must first be estimated. The optimal detector in a single-user context Matched Filter (MF) [8]: this technique is optimal with respect to additive noise in the absence of multiple access interference. It consists in applying on each sub-carrier an equalization coefficient equal to the conjugate complex of the coefficient of the channel. Another detection technique is the Equal Gain Combining Detector (EGC) [4] technique, based on the correction of the phase distortion introduced by the channel. The Orthogonality Restoring Combining detector (ORC) [7] technique makes it possible to cancel the dispersion provided by the channel. As a result of the application of Wiener filtering [11], the technique of Minimum Mean Square Error (MMSE) [5] makes a compromise between minimizing multiple access interference and maximizing Signal-to-Noise Ratio. Hence, the purpose of the MMSE technique is to minimize the value of the mean squared error for each subcarrier between the transmitted signal and the output signal of the detection.

In this paper, a new channel estimation technique is proposed. It is based on the choice of the most appropriate model for the positioning of the pilot carrier, which is considered as a research subject in this field. Interleaved partitions are chosen with the unit signal generator where the pilot carriers are equally organized, in the sense that, successive pilots have equal frequency distances. The choice of a unit signal generator is to suspend the calculations involved in the channel estimation. Theoretically, the received signal is a convolution product between the original signal and the channel coefficients to be estimated. In this way, these coefficients can be estimated in the time domain directly as interleaved partitions represent the best solution for this purpose. The principle of this technique depends on the choice of the position of pilot's carriers in an appropriate manner and then the injection these carriers by the unit signal. Given that, the Inverse Fast Fourier Transform (IFFT) of the unit signal is a pulse signal, this signal can be separated from the time signal of data. In order words, the convolution between the pulse signal and the channel coefficients gives the same coefficients, which means that the time information signal and the channel coefficients can be separated. Simulation results were performed in Rayleigh fading channel with a comparison between the channel estimation and the ideal case. They showed the good performance, the proposed method estimation provides, which approximates the ideal situation.

The remaining body of the paper is organized as follows. MC-CDMA modulation is briefly described in Sect. 2. In Sect. 3, the proposed New Channel Estimation scheme is detailed. The simulation results are shown in Sect. 4. At the end, the section moves to the general conclusion about this paper in its fifth part.

2   MC-CDMA Modulation

2.1   The Transmitter

The MC-CDMA technique [2] is based on the concatenation of spread spectrum and multi-carrier modulation. The MC-CDMA modulator spreads the data of each user in the frequency domain. More specifically, the complex symbol d_i specific to each user i is first multiplied by each of the chips $C_i[k]$ of the spreading code W_{L_c}, then applied to the input of the multicarrier modulator. Each subcarrier transmits an item of information multiplied by a chip of the code specific to this subcarrier. By setting $\alpha = N/L_c$, each user exploits N subcarriers to transmit α data per MC-CDMA symbol. The signal X for the CDMA symbol is expressed as:

$$X_{(b)}(m) = \sum_{i=1}^{N_u} d_i\left[b\alpha + l\right] C_i\left[K\right] \tag{1}$$

where, $m = \{k + L_c.l/k = 0 \text{ to } L_c - 1 \text{ and } l = 0 \text{ to } \alpha - 1\}$. The transmitted signal for the MC-CDMA symbol is:

$$z_{(b)}(t) = \sum_{m=0}^{N-1} X_{(b)}(m)e^{2\pi jf_m t} \tag{2}$$

where, $f_m = \frac{m}{T_s}$. The frequency band B occupied by the main lobes of the subcarriers is equal to

$$B = \frac{N}{T_s} \tag{3}$$

where, T_s is the symbol duration MC-CDMA. The complex envelope $z_{(b)}(t)$ of the modulated signal, sampled at T_s/N, is therefore equal to:

$$z_{(b)}\left(\frac{nT_s}{N}\right) = \sum_{m=0}^{N-1} X_{(b)}(m)e^{2j\pi n\frac{m}{N}} \tag{4}$$

We replace Eq. (1) in Eq. (4) we obtain:

$$z_{(b)}\left(\frac{nT_s}{N}\right) = z_{(b)}(n)$$
$$= \sum_{i=1}^{N_u} \sum_{l=0}^{\alpha-1} \sum_{k=0}^{L_c-1} d_i\left[b\alpha + l\right] C_i\left[k\right] e^{2j\pi n \frac{(k+L_c.l)}{N}} \tag{5}$$

After simplification, the expression of signal $z_{(b)}(n)$ is obtained by

$$z_{(b)}(n) = \sum_{i=1}^{N_u} D_{i(b)}(n).\lambda_i(n) \tag{6}$$

where, $D_{i,(b)}(n) = \sum_{l=0}^{\alpha-1} d_i\left[b\alpha + l\right] e^{2j\pi n \frac{l}{\alpha}}$ and $\lambda_i(n) = \sum_{k=0}^{L_c-1} C_i\left[k\right] e^{2j\pi n \frac{k}{N}}$.

When an infinite sequence of MC-CDMA symbols is transmitted, the output signal of the transmitter is a juxtaposition of the symbols MC-CDMA:

$$z(n) = \sum_{b=-\infty}^{b=+\infty} z_{(b)}(n) \tag{7}$$

2.2 The Receiver

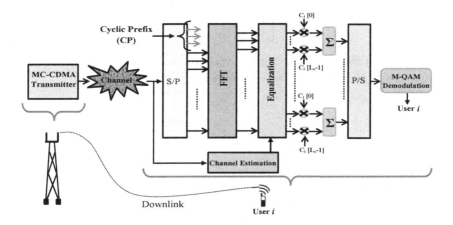

Fig. 1. Block diagram of MC-CDMA receiver

Figure 1 shows the MC-CDMA receiver of the ith user. The MC-CDMA signal received at the input of the receiver is noted as $r(t)$ and is written as:

$$r(t) = (h * z)(t) + w(t)$$
$$r(t) = \sum_{p=0}^{P-1} h_p z(t - \tau_p) + w(t) \tag{8}$$

wher, h represents the impulse response of the channel and P the number of paths discernible by the receiver.

More or less complex equalization techniques associated with the processing of diversity must be implemented at the receiver in order to obtain a correct estimation, after filtering operations, baseband transposition, sampling, delete the guard interval, and the Fast Fourier Transform (FFT) operation the expression of the received signal can be written:

$$r = H.X + w \tag{9}$$

where, r denotes a vector consisting of the values received on each subcarrier:

$$r = [r_0 \ \ldots \ r_{N-1}] \tag{10}$$

and

$$X = [X_0 \ \ldots \ X_{N-1}] \tag{11}$$

The matrix H represents the complex coefficients of the channel of size $M \times N$ The hypotheses previously formulated on the good sizing of the system make it possible to consider this matrix as diagonal:

$$H = \begin{bmatrix} H_0 & 0 & \ldots & 0 \\ 0 & H_1 & \ldots & 0 \\ \vdots & \vdots & \ddots & \vdots \\ 0 & 0 & 0 & H_{N-1} \end{bmatrix} \tag{12}$$

The vector w represents the N components of the noise affecting each subcarrier and model able as so many additives Gaussian processes:

$$w = [w_0 \ \ldots \ w_{N-1}] \tag{13}$$

2.3 Single-User Equalization

Single-user equalization considers only the active user's signal, the other users are considered as jammers. Classically-encountered single-user equalization uses a linear structure, consisting of a one-tap equalizer. Using the preceding matrix notation, it is possible to express Q, the diagonal matrix composed of equalization coefficients q_k:

$$Q = \begin{bmatrix} q_0 & 0 & \ldots & 0 \\ 0 & q_1 & \ldots & 0 \\ \vdots & \vdots & \ddots & \vdots \\ 0 & 0 & 0 & q_{N-1} \end{bmatrix} \tag{14}$$

After equalization and spreading according to the sequence C_i of the user considered, the estimate d_i of the transmitted symbol can be expressed by:

$$\widehat{d_l} = C_i.Q.r \tag{15}$$

where, the matrix C_i represents the spreading codes of size $\alpha \times N$:

$$C_i = \begin{bmatrix} c_j [0 : L_c - 1] & 0 & \cdots & 0 \\ 0 & c_j [0 : L_c - 1] & \cdots & 0 \\ \vdots & \vdots & \ddots & \vdots \\ 0 & 0 & 0 & c_j [0 : L_c - 1] \end{bmatrix} \quad (16)$$

where, $c_j [0 : L_c - 1] = [c_j [0] \ \cdots \ c_j [L_c - 1]]$.

The coefficient q_k is different depending on the choice of detection, and we distinguish three cases:

The Adapted Filter (MF) [8]. It consists in applying on each sub-carrier an equalization coefficient q_k equal to the conjugate complex of the coefficient of the channel H_k, namely:

$$q_k = H_k^* \quad (17)$$

Orthogonality Restoring Combining (ORC) [7] allows to completely canceling the dispersion provided by the channel. The equalization coefficient applied to each sub-carrier is given by:

$$q_k = \frac{1}{H_k} \quad (18)$$

Minimum Mean Square Error (MMSE) [5]. Offers a compromise between minimizing the term of multiple access interference and maximizing the signal-to-noise ratio. It comes from the application of the Wiener filter. The purpose of calculating the equalization coefficient is to minimize the mean squared error for each subcarrier between the transmitted signal and the equalized signal. This resolution leads to the expression of the coefficients q_k:

$$q_k = \frac{H_k^*}{|H_k|^2 + \dfrac{1}{\gamma_k}} \quad (19)$$

The coefficient γ_k is calculated from the estimation of the signal-to-noise ratio by subcarrier.

3   New Channel Estimation Approach

Channel estimation for multicarrier modulations depends on the demodulation processes implemented, two main techniques are used:

– The differential demodulation assumes a quasi-invariance of the channel over a period of two consecutive symbols. This method relies on the coding of the transition from one symbol to another according to the time and frequency axes. It does not require channel estimation at any point, however, it only applies to phase modulations.

– Coherent demodulation is not limited to the chosen modulation. However, it requires the estimation of the frequency response of the channel on all the sub-carriers. It is based on the interpolation of the channel coefficients at the defined positions of the pilot carriers. The possibility of adapting the distribution pattern of these pilot carriers in the time-frequency domain makes them an interesting solution towards mobile radio channels.

According to the sampling theorem, the frequency spacing $N_f \triangle_f$ between two successive pilot carriers, leads to a good estimate if the following relationships are verified (Fig. 2):

$$N_f \triangle_f \leq \frac{1}{\tau_{max}} \tag{20}$$

where, $\tau_{max} = max \{\tau_0, \tau_1, \ldots, \tau_{P-1}\}$ and $\triangle_f = \frac{1}{T_s}$.

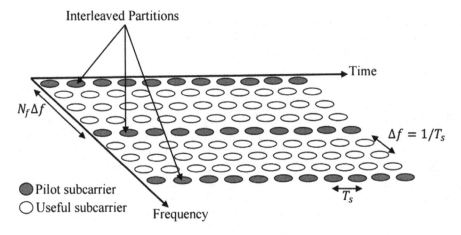

Fig. 2. Representation in the time-frequency domain of the arrangement of the pilot carriers within the frame of MC-CDMA symbol

We can write the transmitted signal in case injection the carriers of pilots by the unit signal as follows

$$z(n) = \sum_{m=1, m \neq l.N_f}^{N-1} X(m)e^{2j\pi n \frac{m}{N}}$$
$$+ \sum_{m=0}^{\frac{N}{N_f}-1} U(m)e^{2j\pi n \frac{N_f m}{N}} / l = 0, 1, \ldots, \frac{N}{N_f} - 1 \tag{21}$$

where, $U(m)$ represents the unit signal, and we have:

$$\sum_{m=0}^{\frac{N}{N_f}-1} U(m)e^{2j\pi n \frac{N_f m}{N}} = \frac{N}{N_f} \sum_{m=0}^{\frac{N}{N_f}-1} \delta \left(n - m \frac{N}{N_f} \right) \tag{22}$$

where, $\delta(n)$ represents a pulse signal, and the signal at the receiver is given by the following relationship:

$$
\begin{aligned}
r(n) &= \sum_{p=0}^{P-1} h_p z(n-\tau_p) + w(n) \\
&= \sum_{p=0}^{P-1} h_p \sum_{m=1, m \neq l.N_f}^{N-1} X(m) e^{2j\pi(n-\tau_p)\frac{m}{N}} \\
&+ \frac{N}{N_f} \sum_{p=0}^{P-1} h_p \sum_{m=0}^{\frac{N}{N_f}-1} \delta\left(n - m\frac{N}{N_f}\right) + w(n)
\end{aligned}
\tag{23}
$$

The channels coefficients can be evaluated directly according to the following relationship:

$$
\widehat{h_p} = \frac{N_f}{N} \sum_{m=0}^{\frac{N}{N_f}-1} r\left(p - m\frac{N}{N_f} mod(N)\right), \; p = 0, 1, \ldots, P-1
\tag{24}
$$

4 Simulation Results

In this section we will evaluate the BER performances. As a simulation environment, we will consider an MC-CDMA system with N = 256 sub carriers and modulating data from a QPSK. For BER evaluation, we will consider a Rayleigh and Rice channels for Additive White Gaussian Noise. Table 1 summarizes the considered MC-CDMA parameters [6].

Table 1. MC-CDMA system sizing parameters

Characteristic parameters	Configuration
Sampling frequency f_s	50 MHz
Number of used subcarriers N	256
Nombres of pilot carriers N_f	32
widthband used W	37.5 MHz
Symbol T_s durations	5.12 μs
Times of the guard interval T_g	0.4 μs
Lengths of L_c spreading codes	16

Table 2 gives the different parameters of the MC-CDMA system for the two channels (Rayleigh, Rice). Note that the size of the guard interval is greater than the maximum delay spread [6].

Table 2. Values of the parameters used for the channels (Rayleigh, Rice)

Channel	Rayleigh	Rice
f_c : Central frequency of the transmitted signal	5.2 GHz	5.2 GHz
B : Widthband of the channel	50 MHz	50 MHz
T_s : Duration of the MC-CDMAsymbol	5.12 μs	5.12 μs
T_g : Duration of the guard interval	0.4 μs	0.4 μs
P : The number of paths	18	18
τ_{max} : The maximum spread of delays	0.2 μs	0.32 μs

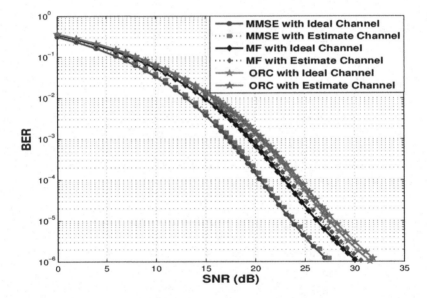

Fig. 3. BER based on SNR, performance of single-user detection techniques over Rayleigh channel

Figures 3 and 4 show the performance of single-user detection techniques for a down-link MC-CDMA system on the Rayleigh and Rice channels, respectively. Where, the modulation used is a QPSK, the spreading codes are the orthogonal codes of Walsh-Hadamard, the detector techniques compared are MMSE, MF and ORC in two cases ideal channel (the receiver knows the coefficients of the transmission channel ideally) and estimated channel (using NCE method). At $BER = 10^{-6}$ the SNR was 27 dB; 30 dB; 32 dB for MMSE, MF, and ORC detection technique in the two cases (ideal channel and estimated channel), respectively. The results obtained for the ORC technique show that this technique fails to remove all MAI (multiple access interference), therefore these performances do not allow to use it in a radio communication context. On the other hand, the MF and MMSE techniques make it possible to restore the orthogonality between the sub-carriers, thus eliminating the term of MAI. Thus, the performances of these techniques are interesting. However, the limited performance of the low-ratio SNR MF technique is due to the noise amplification performed by this technique for the small values of the channel coefficients. Finally, the best performances are obtained by the MMSE technique.

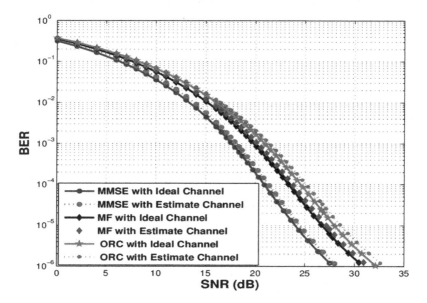

Fig. 4. BER according to SNR, performance of single-user detection techniques on Rice channel

5 Conclusion

In this paper, a New Channel Estimation (NCE) scheme for MC-CDMA system is proposed. MC-CDMA has therefore emerged as an obvious solution to increase the throughput and offer greater robustness compared to the degradation caused by the transmission channel, particularly in urban areas (multi-path channels). In order to improve its performance and quality, we were interested in reserving suitable positions for the pilot carrier, which is mainly based on interleaved partitions. The pilot carrier was also injected by the unit signal. On this basis, the channels have been easily estimated. Simulation results were carried out in different channels, such as Rayleigh and Rice fading channels. The obtained results showed improvements in performance using the proposed model for channel estimation. These performances might closely approximate the results in the ideal case.

References

1. Hanzo, L., Münster, M., Choi, B., Keller, T.: OFDM and MC-CDMA for Broadband Multi-User Communications, WLANs and Broadcasting. Wiley, Hoboken (2005)
2. Kase, Y., Tanaka, M., Seki, T.: Study on characteristics of MC-CDMA communication system. In: 2017 IEEE Asia Pacific Microwave Conference (APMC), pp. 1018–1021 (2017)
3. Kondo, S., Milstein, L.: On the use of multicarrier direct sequence spread spectrum systems. In: Proceedings of MILCOM 1993-IEEE Military Communications Conference, vol. 1, pp. 52–56 (1993)
4. Liu, W., Xu, Z., Yang, L.: SIMO detection schemes for underwater optical wireless communication under turbulence. Photonics Res. **3**(3), 48–53 (2015)
5. Liu, X., Li, Y., Li, X., Xiao, L., Wang, J.: Pilot reuse and interference-aided MMSE detection for D2D underlay massive MIMO. IEEE Trans. Veh. Technol. **66**(4), 3116–3130 (2017)
6. Massiani, A.: Prototyping of high-rate systems based on the combination of multicarrier modulations, spread spectrum technique and multiple antennas modulations. Theses, INSA de Rennes, November 2005. https://tel.archives-ouvertes.fr/tel-00011317
7. Meng, W.X., Sun, S.Y., Chen, H.H., Li, J.Q.: Multi-user interference cancellation in complementary coded CDMA with diversity gain. IEEE Wirel. Commun. Lett. **2**(3), 303–306 (2013)
8. Salahdine, F., El Ghazi, H., Kaabouch, N., Fihri, W.F.: Matched filter detection with dynamic threshold for cognitive radio networks. In: 2015 International Conference on Wireless Networks and Mobile Communications (WINCOM), pp. 1–6. IEEE (2015)
9. Shi, Q., Latva-Aho, M.: Simple spreading code allocation scheme for downlink MC-CDMA. Electron. Lett. **38**(15), 807–809 (2002)
10. Tchiotsop, D., Kengnou Telem, A.N., Wolf, D., Louis-Dorr, V.: Simulation of an optimized technique based on DS-CDMA for simultaneous transmission of multichannel biosignals. Biomed. Eng. Lett. **7**(2), 153–171 (2017). https://doi.org/10.1007/s13534-017-0018-3
11. Yee, N., Linnartz, J.P.: Wiener filtering of multi-carrier CDMA in Rayleigh fading channel. In: 5th IEEE International Symposium on Personal, Indoor and Mobile Radio Communications, Wireless Networks-Catching the Mobile Future, vol. 4, pp. 1344–1347 (1994)

Multi-party WebRTC as a Managed Service over Multi-Operator SDN

Riza Arda Kirmizioglu$^{(\boxtimes)}$ and A. Murat Tekalp

Department of Electrical and Electronics Engineering, Koc University,
34450 Istanbul, Turkey
{rkirmizioglu,mtekalp}@ku.edu.tr

Abstract. We propose a new WebRTC premium service API and a distributed E2E WebRTC service management framework to enable provisioning premium WebRTC services across multiple network service providers (NSP) with software defined network (SDN) infrastructure. The proposed architecture supports both two-party point-to-point (P2P) and multi-party videoconferencing (via a selective forwarding unit (SFU)) with end-to-end (E2E) quality of service (QoS) as a value-added service. Each NSP has full control of its own network resources. We propose a WebRTC service manager (WSM) that is hosted by each NSP. WSMs collaborate with each other in a distributed manner to orchestrate a network slice with specified bitrate and delay over the E2E path between two clients in P2P service and between sending clients and SFU and between the SFU and receiving clients in multi-party service without a need for a central inter-operator service orchestration authority. Clients support scalable VP9 video codec, which is configured according to agreed client upload and download bitrates. Our experimental results show that the proposed managed WebRTC service with guaranteed video quality across multiple NSPs provides lower delay and consumes less network resources compared to today's best-effort WebRTC communications.

Keywords: WebRTC · Scalable video · SDN

1 Introduction

WebRTC is a popular protocol for real-time communications (RTC) that provides browser-to-browser voice, video, and data communications via simple Javascript APIs. VP9 is a royalty-free video codec that allows for efficient scalable video coding. They are supported by the leading browsers such as Firefox and Chrome. WebRTC with VP9 video encoding is currently available over the best-effort Internet. However, RTC services often exhibit undesirable video quality fluctuations due to sudden network bandwidth variations even though

This work was supported by Celtic-Plus project VIRTUOSE and TUBITAK project 115E299.

© Springer Nature Switzerland AG 2020
M. Belkasmi et al. (Eds.): ACOSIS 2019, CCIS 1264, pp. 25–37, 2020.
https://doi.org/10.1007/978-3-030-61143-9_3

WebRTC employs powerful network estimation and video encoding rate adaptation schemes.

Software-defined networking (SDN) is a central theme of the upcoming 5G standard for efficient network slicing to provide desired end-to-end (E2E) quality of service per flow. WebRTC can be offered over SDN with predictable quality.

It is likely that clients in a videoconference have different NSPs possibly at different geographic locations. Hence, this paper proposes an architecture and implementation for third-party video service providers (VSP) to offer WebRTC videoconferencing services at a predictable quality level in collaboration with network service providers (NSP) over a multi-operator SDN environment. The proposed architecture supports both two-party point-to-point (P2P) and multi-party videoconferencing using a selective forwarding unit (SFU).

The main novelties of this paper are:

- We propose a distributed orchestration framework for collaboration between a VSP and multiple NSPs, where each NSP has full control over its network.
- We propose a method for the WebRTC service manager to orchestrate collaboration between multiple network service providers in order to reserve E2E slices between sending clients and the SFU and between the SFU and the receiving clients with bitrates agreed by the clients.
- We propose a method for the WebRTC service manager to set the video encoding rates as well as video upload and download rates for each client given the reserved E2E slice bandwidths.
- We propose a new WebRTC API for a client to request premium service.

The rest of this paper is organized as follows: Sect. 2 reviews the basics of WebRTC and managed video services over SDN. Related works are discussed in Sect. 3. Section 4 presents managed WebRTC services over SDN as a value-added service model over 5G networks. Implementation and evaluation are discussed in Sect. 5. Finally, Sect. 6 presents conclusions.

2   Background

2.1   Overview of WebRTC

WebRTC is an IETF protocol specification together with a number of Javascript API specifications defined in a series of W3C documents [1]. WebRTC provides API for media capture and for sharing media and data streams between browsers. A signaling server is needed for establishing connections between clients and create media paths. Signaling methods and protocols are not specified by WebRTC.

WebRTC packets can traverse network address translation (NAT) gateways, which map a group of private IP addresses to a single public IP address, and firewalls. WebRTC APIs use STUN servers to help end-points find each other in the presence of NATs, and TURN servers as relays if peer-to-peer connections fail in the presence of firewalls.

WebRTC supports the real-time transport protocol (RTP) over the user datagram protocol (UDP). Real-time transport control protocol (RTCP) is used to monitor and estimate the available bandwidth between end systems. WebRTC platform employs the Google congestion control algorithm [2] for rate control of the VP9 video encoder.

2.2   Managed Video Services over SDN

Managed video services refer to those services where the end-to-end bitrate and delay can be controlled. Today managed video services are offered over closed networks, such as IPTV solutions.

SDN introduces a new paradigm that separates the control and data planes and centralizes the control function [3]. The controller communicates with all switches by means of a southbound interface, such as OpenFlow, to specify or modify flow tables. SDN provides a practical framework to offer managed video services over open networks since the controller has an overall view of all network resources, e.g., switch/queue states and traffic load.

Although video traffic is increasing continuously over the years, this is not contributing to the revenues of the NSPs. The advent of WebRTC services combined with NSPs deploying SDN, especially to manage their edge-access networks, provides an important opportunity for third-party VSPs and NSPs to offer managed real-time communications services to increase their revenues.

3   Related Works

A proprietary multi-party videoconferencing system using scalable video coding was first proposed in [4] in order to avoid transcoding at the multi-point control unit (MCU). Scalable coding is now a mainstream technology in multi-party videoconferencing, where a central selective forwarding unit (SFU) selectively forwards layer streams received from sending clients according to terminal type and connection bandwidth of receiving clients.

Our previous works on best-effort WebRTC videoconferencing with scalable video coding include motion-adaptive resolution layer selection at clients for two-party point-to-point conferencing [5] and motion-adaptive rate selection at clients and motion-adaptive layer selection at the SFU for multi-party conferencing [6]. Other best-effort multi-party WebRTC videoconferencing solutions include P2P-MCU architecture, where the MCU functionality is hosted in one of the client peers [7] and implementation of the MCU or SFU in the cloud [8].

We previously proposed managed DASH-unidirectional video streaming services over SDN [9]. However, prior work on managed WebRTC videoconferencing services over SDN is very limited. Launching WebRTC services in a single-operator SDN environment was discussed in a concept paper on "network as a service" [10], which did not have an implementation. Implementation of dynamic-network-enabled RTC on a proof-of-concept 5G network was discussed in [11]. Our previous work on managed multi-party WebRTC services covers the case of mesh-connected clients over single-domain SDN [6]. We also proposed a method for service level agreement (SLA) negotiation between multiple SDN operators for managed E2E video services [12]. This paper proposes, for the first time, a managed multi-party WebRTC architecture using an SFU over a multi-operator SDN environment, which extends [12] and [6].

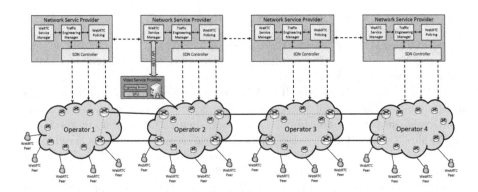

Fig. 1. Architecture for E2E managed WebRTC service provided by a third-party VSP (blue) in collaboration with multiple NSPs with SDN infrastructure. The primary NSP in this case is Operator 2 since the WebRTC VSP servers (Signalling Server and SFU) are connected to the network of NSP2. (Color figure online)

4   Multi-operator Managed WebRTC Service over Reserved E2E Network Slices

We introduce a framework for third-party VSP to offer managed multi-party WebRTC videoconferencing services. Section 4.1 presents the proposed service architecture. Sections 4.2 discusses intra-domain path computation and sharing aggregated domain graph by individual NSPs. Section 4.3 describes E2E service orchestration by the primary NSP and QoS queue management by other NSPs.

4.1   Multi-operator Managed WebRTC Service Architecture

A traditional RTC video service provider (VSP) does not collaborate with the NSP; thus, offers a best effort service with no control over the network. In the proposed managed WebRTC services, the VSP collaborates with the NSP, where the NSP provides network slices to offer per-flow end-to-end quality of service (QoS) by computing paths between end points, with specified bandwidth and delay parameters, and performing queue management at switches.

An architecture that enables a third-party VSP to provide managed WebRTC services in a multi-operator SDN environment is depicted in Fig. 1. The key elements of the proposed architecture are a *WebRTC VSP* capable of providing such service and a *WebRTC service manager* (WSM) at each NSP to orchestrate network functions to enable such service. We call the NSP to which the VSP servers are connected as the *"primary NSP"*, e.g., Operator 2 in Fig. 1.

We assume that each NSP hosts a *traffic engineering manager* (TEM) to manage the traffic over its own network domain. The functionality of these modules is discussed in the following:

WebRTC VSP: WebRTC VSP is a third party entity running a *WebRTC signaling server* (WSS) and a *selective forwarding unit* (SFU) to support multiparty service. Peers connect to the WSS to establish a session by exchanging Session Description Protocol (SDP) and Internet Connectivity Establishment (ICE) objects. The ICE object contains IP address, transport protocol and port number of a peer. WSS creates data sockets for peers, and also passes peer information to the WSM of the primary NSP via the REST API.

WebRTC Service Manager (WSM): Each NSP hosts a WSM as a northbound SDN controller application to manage WebRTC traffic of a specific VSP. The WSM of the primary NSP receives peer information for sessions from the WSS through REST API, and shares them with WSM of other NSPs. The WSM of the primary NSP is also responsible to communicate computed reserved slice bandwidths and other session information received from TEM to the WSS, which passes them back to peers via already created data sockets.

Traffic Engineering Manager (TEM): Each NSP hosts a TEM to manage the traffic over its network. In addition, the TEM of the primary NSP is responsible to orchestra E2E reserved slices as discussed in Section 4.2. Once all paths are computed, each TEM passes relevant path information to its SDN controller, and the controller updates flow tables of switches in its domain and manages their queues accordingly.

WebRTC Policing Unit (WPU): WPU is responsible to enforce that every WebRTC flow does not exceed the reserved bandwidth in order to elicit fairness.

4.2   Virtual Path Computation by Each NSP

We classify NSPs as one of primary NSP, source NSP, target NSP, or a transient NSP. The TEM of each NSP performs path computation and queue management tasks between the source and destination switches within its domain since each NSP has full autonomy to control the traffic over its network. The source switch can be a switch that a sending peer is connected (for source NSP) to or a set of entry border gateway switches (for a transient NSP). The destination switch can be a switch that a receiving peer is connected to (for a target NSP) or a set of exit border gateway switches (for a transient NSP).

Since the SDN controller of each NSP has full visibility of traffic and switch queue states within its domain, the TEM of each NSP (other than the primary NSP) independently computes a path between all possible source and destination switches in its domain that satisfies a set of bandwidth and/or maximum delay requirements using Algorithm 1. Note that path computation at each NSP is dynamic, i.e., it is repeated periodically in a dynamic network environment.

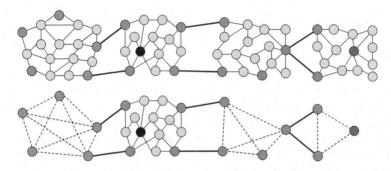

Fig. 2. The actual multi-operator data plane network and its aggregated graph representation as seen by operator (NSP) 2. Blue switches are border gateways of NSPs. WebRTC peers are connected to black and red switches. (Color figure online)

Algorithm 1: Constrained Shortest-Delay Path Finder

---

1 **Input:** G, $sender_i$, $receiver_i$, bw_i
2 **Output:** p_i^*, $delay_i$
3 $p_i^* \leftarrow \emptyset$ (Set of links)
4 $delay_{p^*} = 0$
5 Create graph G^* such that $\forall \text{Link} \in G^* \geq bw_i$
6 $p_i^* \leftarrow \text{Dijkstra}(G^*, sender_i, receiver_i)$
7 $delay_{p^*} = $ Sum of all latency delay over p_i^*

Inputs to Algorithm 1 are a directed graph G that represents the complete topology of a particular NSP, the source switch $sender_i$, the destination switch $receiver_i$, and the desired bandwidth bw_i. The outputs are the optimal path p_i and delay $delay_i$ calculated by the NSP.

Complexity Analysis: The complexity of creating the graph G^* with adjacency list data structure is $O(|E| + |V|)$ and running the Dijkstra's Shortest Path algorithm [13] with binary heap is $O(|E|log|V|)$. Therefore, the worst-case complexity of Algorithm 1 is $O\big(|E| + |V| + |E|log|V|\big)$.

The TEM of each NSP (other than the primary NSP) passes all computed virtual paths to its WSM which passes this information to the WSM of the primary NSP in the form of a aggregated domain graph, so that the TEM of the primary NSP can perform E2E path computations based on this aggregated graph as discussed in Sect. 4.2. Then, the TEM of the primary NSP has an aggregated view of the entire network in the form of all possible virtual paths between border gateways of all NSPs. An example aggregated view of the entire network as seen from the primary NSP is shown in Fig. 2.

4.3   Multi-operator Managed WebRTC Service Provisioning

This section provides a step-by-step description of E2E managed WebRTC service provisioning over multiple NSP, which consists of session initiation and end-to-end network slice orchestration by the primary NSP.

Session Initialization. Clients initiate a premium session by signing in to the WSS and exchanging SDP and ICE objects as they normally do to initiate a WebRTC session. SDP includes client terminal type with their video capture/display resolutions. The ICE object contains the client IP address, transport protocol and port number. In a premium service, WSS passes these objects to the WSM of the primary NSP, which collaborates with the WSM of other NSPs to initiate the E2E inter-operator slice reservation work flow.

End-to-End Network Slice Orchestration. Once the primary WSM gathers session information from WSS, it instantiates relevant objects and shares them with WSM of other NSPs. Other WSMs share aggregated graph information including delay metrics with the primary WSM. The source and target NSPs mark the switch that the source and target peers are connected to in their aggregated sub-graphs, respectively. Note that the aggregated sub-graphs can contain multiple paths between source and destination switches with base and enhancement layer bitrates since peers may desire to receive base layer only video or base and enhancement layer video, which require different bitrates. The primary WSP passes the aggregated view of the entire network to its TEM, which then computes E2E paths over the aggregated graph using Algorithm 2.

Algorithm 2 searches for a path with a bandwidth bw_i between every pair of sender($sender_i$)-SFU($receiver_i$) and SFU($sender_i$) - receiver($receiver_i$). In line 3 it determines the best possible intermediate domains by running Algorithm 1 over aggregated graph G_{aggr} for a particular $sender_i$ - $receiver_i$ pair. The main NSP creates source (src_{list}), destination (dst_{list}) and domain (C_{list}) lists in line 4–7. In line 8 $sender_i$ is added to src_{list}. Then, in line 9–16 for each $Link$ in aggregated path p_{agg}, if $Link$ is inter-domain its source and destination nodes added to src_{list} and dst_{list} in reverse manner. Since, Inter-domain link's source node is the destination of the node $sender_i$ and inter-domain link's destination node is next source node for the domain where it resides. If the $Link$ is intra-domain (virtual) link, its domain c_j is found and added to C_{list}. The final destination node to which receiver unit connected is added to dst_{list}. The rest of the algorithm runs in a distributed manner. In line 18–23 each NSP c_j runs Algorithm 1 to compute shortest paths and make bandwidth reservation between source node src_j and destination node dst_j.

Complexity Analysis: The complexity of Algorithm 1 was found to be $O(Z)$, where $Z = (|E| + |V| + |E|log|V|)$. Assuming there E edges in aggregated graph and V nodes. Line 8–17 runs in $O(Z + |E| + |V|)$. If the number of operators are said to be C, line 18–21 runs in $O(CZ)$ and line 22–23 runs in $O(|E| + |V|)$.

Algorithm 2: Distributed End-to-End Path Computation

---

1 **Input:** G_{aggr}, $sender_i$, $receiver_i$, bw_i

2 **Output:** p_i, $delay^*$

3 $p_{agg} \leftarrow$ CSPF(G_{aggr}, $sender_i$, $receiver_i$, bw_i)

4 $C_{list} \leftarrow \emptyset$ (Set of controller domains)

5 $src_{list} \leftarrow \emptyset$ (Set of Entrance Gateways)

6 $dst_{list} \leftarrow \emptyset$ (Set of Exit Gateways)

7 $delay^* = 0$

8 src_{list}.add($sender_i$)

9 **foreach** $Link \in p_{agg}$ **do**

10   **if** $Link$ is Inter-Domain **then**

11   src_{list}.add($Link$.destinationNode)

12   dst_{list}.add($Link$.sourceNode)

13   **if** $Link$ is Intra-Domain **then**

14   Find Controller c_j that $Link$ belongs to

15   **if** c_j is not an element of C_{list} **then**

16   C_{list}.add(c_j)

17 dst_{list}.add($receiver_i$)

18 **foreach** $c_j \in C_{list}$, $src_j \in src_{list}$, $dst_j \in dst_{list}$ **do**

19   p_j, $delay_j \leftarrow$ CSPF(c_j.Graph, src_j, dst_j, bw_i)

20   p_i.add(p_j)

21     $delay^* = delay^* + delay_j$

22 **foreach** $c_j \in C_{list}$, $p_j \in p_i$ **do**

23   Reserve Bandwidth bw_i on p_j in c_j

However, line 18–21 runs for each different NSP in parallel; therefore, the total computational complexity of Algorithm 2 is $O(|E| + |V| + |E|log|V|)$.

The TEM passes the computed E2E path to its WSM. The WSM of the primary NSP then informs other domain WSMs about the flow within respective NSPs and makes final path reservations in all other NSP networks.

Queue Management at Switches by Each NSP. The bandwidth reservation within each NSP domain along the E2E path is managed by the SDN controller of the respective NSP. To this effect, the SDN controller of each NSP along the reserved E2E path sets up a special priority queue, whose resources are updated dynamically depending on cumulative needs of all managed flows, on each switch along the path and all managed WebRTC flows are directed to this priority queue. When a new WebRTC flow is initiated the resources of affected queues are incremented, and when a WebRTC flow expires, the bitrate limit of affected queues are reduced accordingly.

5   Evaluation

We first discuss emulation tests in Sect. 5.1. Simulation results to evaluate the performance of the proposed multi-operator service are provided in Sect. 5.2.

5.1   Emulation Tests

We first emulated the proposed architecture to demonstrate that the proposed service successfully runs over real network protocols using a WebRTC signaling server, an SFU, and three native WebRTC peer applications as a proof of concept, where each peer runs in a separate personal computer.

We emulated a multi-operator network with three NSP domains using the Mininet [14], where each NSP domain has 15 open virtual switches (OVS). Each NSP has one border gateway, where border gateways are connected to each other using GRE tunnels. Each NSP domain runs a separate instance of the Floodlight controller [15], which communicates with the switches in its domain through its southbound OpenFlow interface. The video service provider (VSP), which runs a WebRTC signaling server (WSS), and a Janus SFU [16], is a virtual host in one of the NSP domains. The WSS gathers IP addresses and terminal types of all WebRTC peers that are signed in. The Janus application runs on the same virtual host that runs the WSS application.

There are three peers that run WebRTC in Chrome browsers [17] on real hosts (three separate laptops), one in each NSP domain, which are connected to the server running the Mininet using ethernet to USB hardware interfaces (hw-intf). We use real hosts to run WebRTC in Chrome browsers on separate laptops rather than on virtual hosts because running the Mininet, SDN controllers, WebRTC signaling server, and multiple instances of VP9 encoder/decoder on the same server causes real-time CPU performance limitations. The WebRTC peers are directed to Janus Gateway VP9-SVC video room, which serves as the SFU.

5.2   Simulation Results

Since it is difficult to build a large scale emulation environment, we provide performance results using a large scale simulation environment with tens of NSP domains, thousands of switches, and hundreds of WebRTC sessions involving thousands of peers.

We simulated 16 NSP domains with the inter-domain topology shown in Fig. 3, where each NSP is connected to each of its neighbor NSPs through 2 border gateways. We simulated different 16-NSP networks of different sizes, where each NSP has 25, 36, 49, ..., 169 switches (nodes) with random topology. We employ a physical distance based probabilistic model [18] to create bi-directional links between pairs of nodes within each NSP. The probability P_{ij} of a link between two nodes i and j is given by

$$P_{ij} = \beta \exp\left(\frac{-d_{ij}}{L\alpha}\right)$$

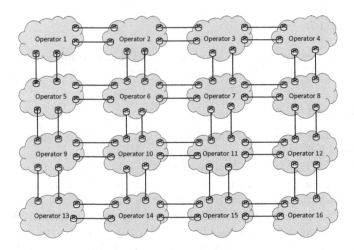

Fig. 3. Inter-domain topology of simulated 16-NSP network.

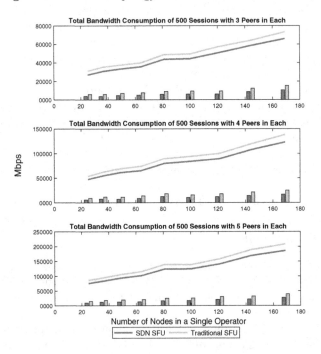

Fig. 4. Total E2E bandwidth consumption of 500 WebRTC services.

where d_{ij} is the Euclidean distance between the nodes, L is the maximum Euclidean distance between any two nodes, and α and β are parameters. In order to model distances between nodes, we consider a rectangular grid of unit squares, where there are as many squares as the number of switches in an NSP.

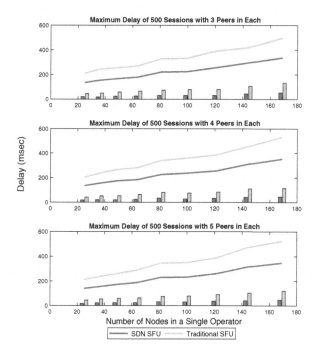

Fig. 5. Maximum E2E delay of 500 WebRTC services.

Each switch is placed at a random coordinate within a unit square. We assume that each link introduces a fix delay of 5 ms, which is determined as the average of *Traceroute* measurements [19] from our server to different university servers.

We compare the performance of the proposed WebRTC service architecture with that of the best-effort service in terms of i) the total bandwidth consumption of WebRTC service users that is of interest to NSPs, and ii) the average and maximum E2E delay of all WebRTC sessions that are indicators of service quality relevant to VSP and end users.

For each network size, we ran 3 experiments, where there are 500 WebRTC sessions with 3, 4, and 5 peers in each session, respectively. We repeat each experiment 100 times with random switch positions and links in each NSP, peers connected to random switches, random location of the central SFU server, and randomly selected video resolution for each peer. The switch that each peer and the central SFU server is connected to is chosen according to a uniform PMF; i.e., we draw a uniform random number between 0 and 1, and multiply this number with the total number of switches to determine a randomly selected switch.

In the managed multi-operator service, routing is computed using Algorithm 2. In the best-effort service, inter-domain routing is performed via pre-determined routes using the nearest border gateway. In our implementation the best-effort traffic follows the nearest city block distance first along the columns and then along the rows referring to Fig. 3.

The average and standard deviation of total bandwidth and maximum delay in 500 sessions with 3, 4, and 5 peers per session vs. network size are shown in Fig. 4 and Fig. 5, respectively. Results show that the proposed managed WebRTC service offers value to both NSPs and end users since it consumes less overall network resources and provides better E2E delay performance compared to the best-effort WebRTC service. This observation is consistent over all network sizes.

6   Conclusions

We propose a distributed managed multi-operator WebRTC service architecture and algorithms for a third-party VSP to provide such services over reserved network slices in collaboration with multiple NSPs (operators). Since different peers can be in different geographical locations in a videoconference and have different NSPs, the proposed architecture includes a WebRTC service manager that orchestrates E2E network slice provisioning across multiple NSPs. We implemented an emulation environment using the Mininet with real OVS and Floodlight SDN controllers for each NSP domain to validate the proposed architecture. Communications between the controllers of NSP domains is realized using the protocol that we developed in our earlier work [12]. Our simulation results clearly demonstrate that the proposed managed WebRTC service with guaranteed video quality provides lower delay and consumes less network resources compared to today's best-effort WebRTC communications.

References

1. Alvestrand, H.: Overview: real time protocols for browser-based applications. draft-ietf-rtcweb-overview-19 (2017)
2. Carlucci, G., De Cicco, L., Holmer, S., Mascolo, S.: Analysis and design of the Google congestion control for web real-time communication (WebRTC). In: Proceedings of the 7th International Conference on Multimedia Systems (2016)
3. Software-defined networking: The new norm for networks, Open Networking Fundation (ONF). White Paper (2012)
4. Eleftheriadis, A., Civanlar, M.R., Shapiro, O.: Multipoint videoconferencing with scalable video coding. J. Zhejiang Univ. Sci. A **7**(5), 696–705 (2006). https://doi.org/10.1631/jzus.2006.A0696
5. Bakar, G., Kirmizioglu, R.A., Tekalp, A.M.: Motion-based adaptive streaming in WebRTC using spatio-temporal scalable VP9 video coding. In: IEEE Globecom, Singapore, December 2017
6. Kirmizioglu, R.A., Kaya, B.C., Tekalp, A.M.: Multi-party WebRTC videoconferencing using scalable VP9 video: from best-effort over-the-top to managed value-added services. In: IEEE International Conference on Multimedia and Expo (ICME), San Diego, CA, USA, July 2018
7. Ng, K.-F., Ching, M.-Y., Liu, Y., Cai, T., Li, L., Chou, W.: A P2P-MCU approach to multi-party video conference with WebRTC. Int. J. Future Comput. Commun. **3**(5), 319 (2014)

8. Yoon, S., Na, T., Ryu, H.-Y.: An implementation of Web-RTC based audio/video conferencing system on virtualized cloud. In: IEEE International Conference on Consumer Electronics (ICCE) (2016)
9. Bagci, K.T., Sahin, K.E., Tekalp, A.M.: Compete or collaborate: architectures for collaborative DASH video over future networks. IEEE Trans. Multimed. **19**(10), 2152–2165 (2017)
10. Boubendir, A., Bertin, E., Simoni, N.: Network as-a-service: the WebRTC case: how SDN and NFV set a solid Telco-OTT groundwork. In: International Conference on the Network of the Future (NOF), Montreal, Canada, 30 September–2 October 2015
11. Jero, S., Gurbani, V.K., Miller, R., Cilli, B., Payette, C., Sharma, S.: Dynamic control of real-time communication using SDN: a case study of a 5G end-to-end service. In: IEEE/IFIP Network Operations and Management Symposium (NOMS) (2016)
12. Bagci, K.T., Yilmaz, S., Sahin, K.E., Tekalp, A.M.: Dynamic end-to-end service-level negotiation over multi-domain software defined networks. In: IEEE International Conference on Communications and Electronics (ICCE), Ha Long Bay, Vietnam, July 2016
13. Dijkstra, E.W.: A note on two problems in connexion with graphs. Numerische Mathematik **1**, 269–271 (1959)
14. Mininet Virtual Network. http://mininet.org/
15. Project Floodlight Open-Source SDN Controller. http://www.projectfloodlight.org/floodlight/
16. Janus Gateway. https://github.com/meetecho/janus-gateway
17. WebRTC source code. https://chromium.googlesource.com/external/webrtc/
18. Zegura, E., Calvert, K., Bhattacharjee, S.: How to model an internetwork. In: Proceedings of IEEE Infocom, vol. 2, pp. 594–602, March 1996
19. Malkin, G.: Traceroute Using an IP Option. RFC 1393, January 1993

NCBP: Network Coding Based Protocol for Recovering Lost Packets in the Internet of Things

El Miloud Ar-Reyouchi[1,2]($\boxtimes$) (iD), Yousra Lamrani[3], Imane Benchaib[3], Salma Rattal[4], and Kamal Ghoumid[3] (iD)

[1] Abdelmalek Essaâdi University, 93000 Tetouan, Morocco
e.arreyouchi@m.ieice.org
[2] SNRT Rabat, Rabat, Morocco
[3] ENSAO, University Mohammed First, 60000 Oujda, Morocco
[4] FSTM, Hassan II University, 20000 Casablanca, Morocco

Abstract. This paper proposes a novel network coding-based protocol (NCBP) to manage the state of transmission for recovering lost packets and correct errors to communicate effectively in the Internet of Things (IoT). We present a random network coding (RNC) based scheme for an IoT scenario where one master node needs to communicate with multiple monitor nodes over a wireless channel. We discuss the strength and weaknesses of the two most popular versions of error correction protocol in the literature, Forward Error Correction (FEC) and Automatic Repeat ReQuest (ARQ). We compare them with NCBP to evaluate the performance of the proposed protocol. The results show that the intended error correction protocol can effectively improve the quality of data transmission, recovering lost data, and increasing throughput. Finally, the NCBP is compared to the classical approach and to preserve beneficial throughput performance.

Keywords: Network coding · Wireless network communications · FEC · ARQ · Internet of Thing

1 Introduction

A simple combination technique, given by network coding (NC) [1], can provide many potential gains to the IoT network. It improves network delay metrics [2], optimizes wireless communication systems [3], increases throughput, minimizes forwarding delay, and gives many benefic to network, as shown in [4]. RLNC [5, 6] is a potent NC process that allows network nodes to produce separately and a haphazardly linear combination of input source data into coded packets over a finite field. At the destination, RLNC can decode and reduce the required retransmission. The most used approaches for repairing the missing packets and correcting them in the wireless network are classified into two distinct mechanisms. The authors in [7, 8] evaluate the transport protocols performance of FEC, ARQ, and Hybrid ARQ (HARQ) for the data transmission reliability.

© Springer Nature Switzerland AG 2020
M. Belkasmi et al. (Eds.): ACOSIS 2019, CCIS 1264, pp. 38–49, 2020.
https://doi.org/10.1007/978-3-030-61143-9_4

They consider a limited number of authorized retransmissions. But do not exanimate and determinate their performance.

In a wireless network connection, the FEC technique allows the IoT [9] to provide a perfect platform to guarantee a reliable wireless connection. In broadcasting and simplex mode, FEC is widely used in Rayleigh fading channels, and it is employed for detecting and correcting the errors. Without a reverse channel, it is the most used, but also he is frequently employed in the internet, IoT, and wireless communication [10], where the reversible connection is allowed. For the Internet, FEC restores the missing bit. However, FEC is not the only technique required for the error control scheme. It often combines with another scheme like ARQ, as mentioned in [11], such that minor errors are rectified without the need for retransmission, and significant errors are remedied by way of a retransmission service. The convolutional codes and block codes count among the two top relevant categories of FEC.

The protocol ARQ is based on retransmitting the dropped and missing packets using Acknowledgement (ACK), and negative-acknowledgment (NAK or NACK) for every packet. It is worth remembering that the FEC is based on the introduction of the controlled redundancy into the native message before transmission. Instead of retransmission, this redundancy is exploited to restore any lost, dropping, and missing packets at the receiver. As a result, the basic techniques of these two protocols are engaging in the IoT.

In [12], the authors adopt a new technique that includes word interleaving, FEC, and ARQ to minify the error and loss effects encountered in wireless Internet applications. However, in [13], the authors provide an overview of the IoT with emphasis on other protocols and application issues. When reception conditions become severe, many retransmissions are mandatory to successfully transmit the data, resulting in a significant increase in latency and reduces throughput.

Throughput, packet loss, and forwarding delay are three of the most fundamental measures, which can improve network performance for IoT. NC can improve throughput in all scenarios (unicast, multicast, and broadcast). In multi-rate scenes, NC has shown gain higher than doubling the performance [14]. The proposed NCBP allows the IoT devices to provide a reliable service for mission-critical applications like Supervisory Control and Data Acquisition (SCADA) [15] and Telemetry for renewable energy, Oil & Gas distribution, and many other smart critical uses in the field of IoT.

This paper introduces an RLNC based protocol for recovering lost packets. It addresses the problem of packet loss recovery of packets transmitted from the sink node to the monitor nodes. We show that NC can also significantly reduce the number of retransmissions and recover lost packets, thereby provide gains in IoT applications. We propose a network coding-based protocol for retrieving lost packets employing single retransmission, improving Actual Data (AD) message size [16], and throughput.

This paper is divided into five sections, including the introduction. Section 2 presents the problem statement, proposed a solution, and the system model. Section 3 gives the protocol description. The results are provided in Sect. 4. Finally, Sect. 5 determines the conclusions.

2   Problem and Solution Statement and System Model

In this section, we present the challenger of obtaining an effective means for packet loss recovery in the context of IoT.

2.1   Problem Statement

In reality, FEC allows the IoT devices to detect and correct errors without retransmission; consequently, it can improve the data transmitting efficiency. However, it also requires long error correction code, because it sends the code to the IoT device no matter the code is correct or not, which attach useless overhead in low error cases. FEC significantly minimizes packet loss; in contrast, the delay and throughput became particularly long and low, respectively. Also, the FEC coding has a low coding rate, which implies inefficient exploitation of channels; besides, it cannot cancel or minimize jitter except out-of-order packets, which are common in the Internet network.

The improvement can come specifically at the expense of the AD throughput or of the average delay to transmit a single packet between two IoT nodes. FEC protocol imposes a more considerable bandwidth overhead. It also places a higher computational demand on the receiving device because the redundant information in the transmission must be interpreted according to a predetermined algorithm.

ARQ requires a reverse channel for transmission of ACKs/NAKs; it causes poor signal conditions to result in low transfer speeds, causes delay variations due to retransmitted data. We compare the performance of the proposed protocol NCBP to the three major types of packet-loss recovery techniques ARQ, FEC, and HARQ.

2.2   Proposed Solution

The solution consists in reducing the number of transmissions using RNC [17] for recovering lost and dropped packets. It can effectively improve network performance. The NCBP solution appears relatively straightforward and classic, but in practice, has a very beneficial effect. It is practical and profitable to explore the proposed protocol based on RNC in the Internet of Things. Its performance becomes more potent than the functional exploitation of RNC. RNC technique can provide a great many potential gains to the wireless network, including reducing the retransmission number of recovering packets loss, decreasing latency, and enhancing the robustness of the network topology.

The proposed protocol is designed to protect AD that the packet is delivering to the destination against transmission failures. In the case where a packet is fixed-length, the body or data of a packet may be padded with new information to make it the right length.

2.3   System Model

We propose the case where the sink wants to transmit n packets received from network 1, Fig. 1, to the nodes in network 2. In other words, the sink node acts as a relay between networks 1 and 2. The paper assumes that the sink relays n packets to network 2 and that each node in network 2 can receive just one packet. Assume a monitoring system model where IoT nodes are wirelessly connected and communicated, as represented in Fig. 1. The purpose of this monitoring system model is to provide data that accurately reflects the critical conditions of renewable energy [18, 19].

Consider two wireless networks; each one consists of n nodes (monitoring stations in area 1 and area 2) labeled $N_{1.1}, N_{1.2}, \ldots, N_{1.n}$ and $N_{2.1}, N_{1.2}, \ldots, N_{2.n}$ respectively, as shown in Fig. 1.

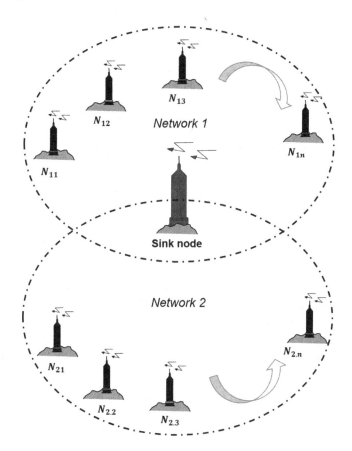

Fig. 1. System model.

The nodes $N_{1.j}, j = 1, 2, 3 \ldots n$, (Network 1) directly transmit their data to the sink node (which is the intermediate node between the two networks).

Assume that every node, in the network 1, transmits one packet, which is $p_{1.1} p_{1.2}, \ldots p_{1.n}$. In a single time step, a sink node can collect all transmitted packets properly from $N_{1.1}, N_{1.2}, \ldots, N_{1.n}$. In general, the research objective describes the data collection by the sink node that sends data to the IoT nodes in wireless network 2 using RNC. IoT sink node has an average of network 1 ($N_{1.1}, N_{1.2}, \ldots, N_{1.n}$) and of network 2 ($N_{2.1}, N_{1.2}, \ldots, N_{2.n}$).

To judge the potency of the proposed protocol, in an error-free transmission, we recall that the efficiency of a protocol $(P_{prot(er=0)})$ is the ratio between the number of serviceable packets transmitted (P_a) to the packets total number transferred (n): $P_{prot(er=0)} = P_a/n$.

Consider P_e, as the probability for erroneous packets that are transmitted, $(1 - P_e)$ is the probability that a packet is correctly transmitted. If the transmission relates to n packets, the possibility for those n packets are successfully sent is: $P_{rob} = (1 - P_e)^n$. According to [1], the protocol efficiency in case of fault is compared without error: $P_{prot(er \neq 0)} = P_{prot(er=0)} \cdot (1 - P_e)^n$.

3 Protocol Description and Simulation Parameters

3.1 Protocol Description

Consider that the topology in Fig. 1 is well respected. The sink node wishes to disseminate n packets, arriving from $N_{1.1}, N_{1.2} \ldots, N_{1.n}$, denoted $p_{1.1}, p_{1.2}, \ldots, p_{1.n}$, to $N_{2.1}, N_{2.2}, \ldots, N_{2.n}$. The $N_{2.1}, N_{2.2}, \ldots, N_{2.n}$ are in the communication range of the sink node and might receive the packets $p_{1.1} p_{1.2}, \ldots, p_{1.n}$. The sink node disseminates the packet $p_{1.1}$, but only the IoT node $N_{2.1}$ receives it. Afterward, the sink node transmits the packet $p_{1.2}$, but at this time, only the IoT node $N_{2.2}$ correctly receives $p_{1.2}$. This process continues until n packets are sent (so forth until packets transmission number is n). Afterward, the sink node transmits the packet $p_{1.n}$, and finally $N_{2.n}$ correctly receives the packet $p_{1.n}$. Thus, each IoT node receives a different packet.

However, the poor propagation conditions require many retransmissions to successfully transmit the data, resulting in a significant increase in latency and a significant decrease in throughput.

The traditional ARQ protocols [20], the store-and-forward approach of one acknowledgment for every packet, makes classical schemes inappropriate for lost packet recovery also for energy conservation over the sink node to mobile IoT node. Also, the retransmission mechanism recovers can detect and recover the lost packets. According to the scenario shown in Fig. 1, the sink node retransmits packets $p_{1.1}$, $p_{1.2}, \ldots, p_{1.n}$. If these retransmissions are successful, $N_{2.1}$ will receive the lost packet $\cancel{p_{1.1}}, p_{1.2}, p_{1.3}, \ldots, p_{1.n}$ and IoT $N_{2.2}$ will receive the missing packet $p_{1.1}, \cancel{p_{1.2}}, p_{1.3}, \cdots$ and $p_{1.n}$. So forth until $N_{2.n}$ will receive the lost packet $p_{1.1} p_{1.2}, \cdots p_{1.n-1}, \cancel{p_{1.n}}$. Thus, even if these $n - 1$ retransmissions are successful, we need a total of $(n - 1) \times n$ retransmissions and n transmissions for all IoT $N_{2.j}$ to successfully receive n packets, which means that we need $n \times n$ transmissions and retransmissions.

With FEC, the redundancy allows the $N_{2.j}, j = 1, 2. \ldots n$ (network 2) to detect a limited number of errors that might happen anywhere in the message packets, and often to remedy these errors without any retransmission protocols. The FEC technique sends the code to the receiver no matter that the code is correct or not, so it needs long and robust error correction code, and the error patterns must be corrected. Consequently, FEC minimizes packet loss at the expense of higher delay and bandwidth.

In the proposed protocol, the sink node, instead of retransmitting duplicate packets $p_{1.2}$, $p_{1.3}$... and $p_{1.n}$, retransmit the coded packet, which takes the following general format $p_{1.1} \oplus p_{1.2} \oplus \ldots \oplus p_{1.n}$.   IoT $N_{2.1}$ node   can   decode   packets $p_{1.2}$, $p_{1.3}$... and $p_{1.n}$, after receiving this packet, by using the XOR operator [21], between the packet $p_{1.j}$, $j = 1, 2, \ldots. n$, which it has already received, over the coded packet. Thus, to obtain the originals packets of the n IoT sources, the $N_{2.1}$ decodes $p_{1.j}$, $j \neq 1$. It applies the XOR operator between the packets $p_{1.2}$, which it has already received   over   the   coded   packet   since $p_{1.j} = p_{1.1} \oplus (p_{1.1} \oplus p_{1.2} \oplus \ldots \oplus p_{1.n})$ where $j \neq 1$. In the same way, the $N_{2.2}$ node can decode $p_{1.j}$, $j \neq 2$ since $p_{1.j} = p_{1.2} \oplus (p_{1.1} \oplus p_{1.2} \oplus \ldots \oplus p_{1.n})$ where $j \neq 2$ by using the XOR operator between the packets $p_{1.2}$, which it has already received, over the coded packet. Finally, $N_{2.n}$ decodes $p_{1.n}$ by applying the XOR operator between the packets $p_{1.n}$ that it has already reached and   correctly   received,   over   the   coded   packet   since $p_{1.j} = p_{1.n} \oplus (p_{1.1} \oplus p_{1.2} \oplus \ldots \oplus p_{1.n})$ where $j \neq n$.

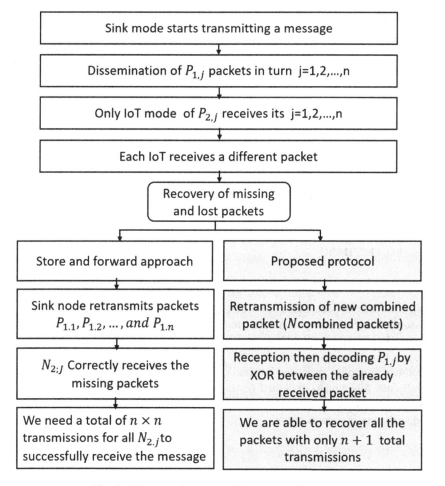

Fig. 2. The operation steps of the proposed protocol.

Thus, we can recuperate all the lost packets with only $n + 1$ total transmissions. This example illustrates the benefit of using Network Coding in a single-hop topology example. The number of transmissions was reduced from $n \times n$ to $n + 1$. Therefore, for $n = 2$, the number of transmissions reduced from 4 to 3, whereas for $n = 10$ it reduced from 100 to 11. Consequently, the number of transmissions and retransmissions are greatly reduced for a large number of IoT nodes. Based on the procedure described above, we can significantly improve many potential gains to the network, including achieving the network capacity, decreasing latency, and improving robustness to network dynamics.

Assume we wish to transmit n error-free packets, and each packet is retransmitted correctly. Figure 2 shows the operation steps of the proposed protocol.

3.2   Simulation Model Parameters

In the system model, Fig. 1, sink node, and receivers are separated by variable Kbit/s links (from 18 Kbit/s to 152 Kbit/s). The packet size again changes (AD increases from 0 bytes to 1500 bytes), and each link can cause a propagation time and processing delay of 10 µs, sequentially. Assume that the sink node initiates directly forwarding after it has received the closing bit of the packet, and the queues are empty. The analytic formulation of the simulation parameters is shown in the following Table 1.

Table 1. Analytic formulation of the simulation parameters.

Parameters	Condition chosen
Maximum number of transmissions & retransmissions	400
Actual Data (AD) length	0–1500 bytes
propagation and processing delay	10 ms
Link capacity	18 Kbit/s to 152 Kbit/s
Number of packets	20
FEC	3/4
ARQ, FEC, HARQ	On/Off

We consider that packet loss can occur, as is the case in IoT wireless transmissions, and assume that if a message is not transmitted effectively, retransmission is reinitiated. This procedure is repeated until a successful transmission occurs. We are assuming that successive transmissions are autonomous. In this paper, Matlab is used to implement the optimization of a proposed method and define the NCBP objective function. Generally, the direct paths exploited between the nodes in network 1 and the nodes in network 2 (via the sink node) (i.e., line-of-sight, LOS) are available; consequently, the Rice distribution is a good option.

4   Results

The results focus on the ability of the proposed protocol. They show that the NCBP decreases the number of retransmissions significantly, improves the AD throughput, and the forwarding delay, which are the most fundamental measures for enhancing the wireless communication performance. The NCBP enhances the detection and correction of a lost packet in IoT Communications. As mentioned in Sect. 2 and described in Sect. 3, NCBP significantly reduces the number of retransmissions from $n \times n$ to $n + 1$, as illustrated in Fig. 3.

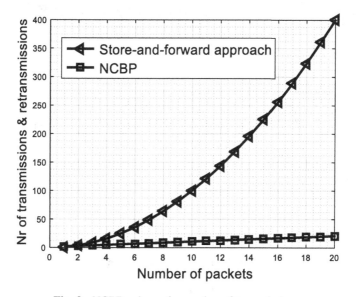

Fig. 3. NCBP reduces the number of transmissions.

With the classical store-and-forward approach, the missing packets are restored, needing a high number of transmissions & retransmissions. While for NCBP, this number is minimal. With 20 packets, the number of transmissions & retransmissions is reduced by 94.75%. The following Table 2 summarizes the percentage results of Transmission & Retransmission number reduction (TRNR) vs. a different number of packets.

Table 2. Percentage of transmission & retransmission number reduction.

Number of packets	5	10	15	20
TRNR		76%	89%	92, 8%	94, 75%

The simulation results show that the number of transmissions can be severely reduced in the wireless network, consequently lost packets are recovered using the most minimum amount of retransmissions.

The results without any error correction (neither FEC nor ARQ technique to detect and correct a limited number of errors), in Fig. 4, show that the throughput increases when the AD length increases. In any error correction protocol, the AD throughput rises significantly from 33 to 137 kbps for AD length until 500 and then increases slowly from 138 to 155 kbps for an AD length ranging from 500 to 1500 bytes.

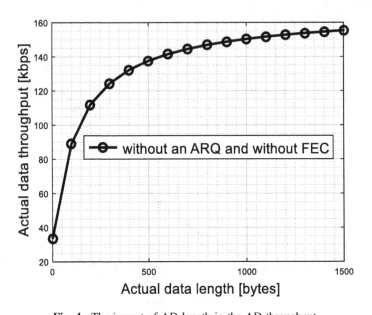

Fig. 4. The impact of AD length in the AD throughput.

In Fig. 5, we compare the proposed NCBP with different types of error correction protocols, namely, FEC, ARQ, and HARQ. With NCBP, the proposed protocol detects and corrects errors with the minimum number of retransmission, unlike the conventional correction, which requires several amounts of retransmission.

It can be observed that throughput without any correction is better than the performance of ARQ, FEC, and HARQ. Furthermore, AD throughput of ARQ protocol is higher than the two other protocols, FEC and HARQ. Consequently, FEC and ARQ significantly reduce the throughput probability of unrecoverable packet loss. Also, Fig. 5 illustrates the impressive improvement in throughput of the proposed protocol NCBP. The following Table 3 summarizes the results of AD throughput (kbps) for different AD length (bytes).

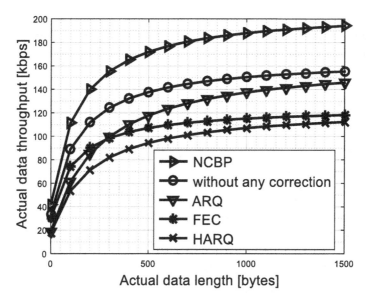

Fig. 5. The Comparison between the NCBP and the three other protocols

Table 3. Actual data throughput for different AD length.

AD length	500	1000	1500
Proposed protocol (NCBP)	171.95	187.92	194.14
Without any error detection and correction protocol	137.56	150.32	155.31
ARQ protocol	117.75	137.35	145.75
FEC protocol	107.14	115.15	118.18
HARQ protocol	94.25	107.08	112.35

In Table 3, the results have compared the ARQ, FEC, and HARQ protocols and the proposed NCBP protocol. AD Throughput (kbps) extracted by the NCBP protocol is much higher than AD throughput of the ARQ, FEC, and HARQ protocols (37%, 63%, 75% for AD length = 1000 bytes) respectively, as shown in Fig. 5 and illustrated in Table 3. We also note that, without any correction, the AD throughput progressively decreases when compared to the presence of ARQ, FEC, and HARQ protocols. Consequently, the simulation results prove that the proposed protocol outperforms ARQ, FEC, and HARQ protocols, respectively.

5   Conclusions

This paper presents a network coding based protocol for detecting and recovering lost packets. This work proposes an NCBP protocol. NCBP provides significant improvements; it is faster with less retransmission, saving bandwidth and energy, maximizes throughput, and minimizes retransmissions number and latency. The comparison results (with ARQ, FEC, and HARQ protocols) shown that the proposed protocol using network coding can reduce the number of retransmissions and increase throughput, indicating a high IoT network efficiency.

References

1. Ahlswede, R., Cai, N., Li, S.Y., Yeung, R.W.: Network information flow. IEEE Trans. Inf. Theory **46**(4), 1204–1216 (2000)
2. Ar-reyouchi, E.M., Hammouti, M., Maslouhi, I., Ghoumid, K.: The internet of things: network delay improvement using network coding. In: ICC 2017 Proceedings of the Second International Conference on Internet of things, Data, and Cloud Computing, Cambridge, United Kingdom. ACM (2017)
3. Hammouti, M., Ar-reyouchi, E.M., Ghoumid, K., Lichioui, A.: Clustering analysis of wireless sensor network based on network coding with low density parity check. Int. J. Adv. Comput. Sci. Appl. (IJACSA) **7**(3), 137–143 (2016)
4. Ho, T., Koetter, R., Médard, M., Karger, D.R., Effros, M.: The benefits of coding over routing in a randomized setting. In: Proceedings of IEEE I. Symposium on Information Theory (2003)
5. Ho, T., Koetter, R., Médard, M., Karger, D.R., Effros, M., Shi, J., Leong, B.: A random linear network coding approach to multicast. IEEE Trans. Inf. Theory **52**(10), 4413–4430 (2006)
6. Barekatain, B., Khezrimotlagh, D., Maarof, M.A., Quintana, A.A., Cabrera, A.T.: GAZELLE: an enhanced random network coding based framework for efficient P2P live video streaming over hybrid WMNs. Wirel. Pers. Commun. **95**(3), 2485–2505 (2016). https://doi.org/10.1007/s11277-016-3930-4
7. Lin, S., Costello, D.J.: Error Control Coding: Fundamentals and applications. Chap. 15. Prentice-Hall, Upper Saddle River (1983)
8. Kotuliaková, K., Šimlaštíková, D., Polec, J.: Analysis of ARQ schemes. Telecommun. Syst. **52**(3), 1677–1682 (2011)
9. Alabady, S.A., Salleh, F.M., Al-Turjmanc, F.: LCPC error correction code for IoT applications. Sustain. Cities Soc. **42**, 663–673 (2018)
10. Ar-Reyouchi, E.M., Ghoumid, K., Ameziane, K., El Mrabet, O.: Performance analysis of round trip time in narrowband RF networks for remote wireless communications. Int. J. Comput. Sci. Inf. Technol. (IJCSIT) **5**(5), 1–20 (2013)
11. Ar-Reyouchi, E.M., Chatei, Y., Ghoumid, K., Lichioui, A.: The powerful combined effect of forward error correction and automatic repeat request to improve the reliability in the wireless communications. In: International Conference on Computational Science and Computational Intelligence (CSCI), Las Vegas, pp. 691–696 (2015)
12. Chen, D., Rong, B., Shayan, N., Bennani, M., Cabral, J., Kadoch, M., Elhakeem, A.K.: Interleaved FEC/ARQ coding for QoS multicast over the internet. Can. J. Electr. Comput. Eng. **29**(3), 159–166 (2004)
13. Al-Fuqaha, A., Guizani, M., Mohammadi, M., Aledhari, M., Ayyash, M.: Internet of things: a survey on enabling technologies, protocols and applications. IEEE Commun. Surv. Tutor. **17**(4), 2347–2376 (2015)
14. Vieira, L.F., Gerla, M., Misra, A.: Fundamental limits on end-to-end throughput of network coding in multi-rate and multicast wireless networks. Comput. Netw. **57**(17), 3267–3275 (2013)

15. Rezai, A., Keshavarzi, P., Moravej, Z.: Key management issue in SCADA networks: a review. Eng. Sci. Technol. Int. J. **20**(1), 354–363 (2017)

16. Rattal, S., Ar Reyouchi, E.M.: An effective practical method for narrowband wireless mesh networks performance, 1 (2019). Article no. 1532

17. Ar-Reyouchi, E.M., Lichioui, A., Rattal, S.: A group cooperative coding mod-el for dense wireless networks. (IJACSA) Int. J. Adv. Comput. Sci. Appl. **10**(7), 367–373 (2019)

18. Hammouti, M., Ar-Reyouchi, E.M., Ghoumid, K.: Power quality command and control systems in wireless renewable energy networks. In: International Renewable and Sustainable Energy Conference (IRSEC), Marrakech, Morocco, pp. 763–769. IEEE (2016)

19. Li, J., Zhou, Y., Liu, Y., Lamont, L.: Performance analysis of multichannel radio link control in MIMO systems. In: Simplot-Ryl, D., Dias de Amorim, M., Giordano, S., Helmy, A. (eds.) ADHOCNETS 2011. LNICSSITE, vol. 89, pp. 106–116. Springer, Heidelberg (2012). https://doi.org/10.1007/978-3-642-29096-1_8

20. Maslouhi, I., Ar Reyouchi, E.M., Ghoumid, K., Kaoutar, B.: Analysis of end-to-end packet delay for Internet of Things in wireless communications. Int. J. Adv. Comput. Sci. Appl. **9**(9), 338–343 (2018)

21. Katti, S., Rahul, H., Hu, W., Katabi, D., Medard, M., Crowcroft, J.: XORs in the air: practical wireless network coding. IEEE/ACM Trans. Netw. **16**(3), 497–510 (2008)

Vehicular Communications

Vehicle Collision Avoidance

Khawla A. Alnajjar$^{(\boxtimes)}$, Noora Abdulrahman, Fatma Mahdi,
and Marwah Alramsi

Department of Electrical Engineering, University of Sharjah,
Sharjah, United Arab Emirates
{kalnajjar,U00034688,U00036990,U00038227}@sharjah.ac.ae

Abstract. Road accidents are considered as one of the major reasons of deaths. The proposed vehicle collision avoidance system aims to increase the traffic efficiency and safety. Through the proposed system, the driver can be alerted to the surrounding hazards by communicating with other drivers. One of the main advantages of the proposed system is to warn the driver if a distance between vehicles is less than the required safety distance or not. Furthermore, in foggy weather, the system is designed to notify the driver of a nearest vehicle or obstacle because the vision will not be clear.

Keywords: Accident · Vehicular communication system ·
Vehicle-to-vehicle collision avoidance (V2V) · Vehicle to-infrastructure
(V2I) communications

1 Introduction

The rising rate of car accidents all over the world has made this issue to spread in the newspaper and in the social media posts. In the united Arab Emirates (UAE), some of the statistics on 2013 show that road accidents have become one of the major reasons of deaths in the whole country [1]. Accidents are caused by either the drivers' faults or by other factors that the drivers are not guilty about. Lack of safety distance is an example of drivers' mistakes that may lead to accidents. While water accumulation, fog and stray animals are other types of hazards that may cause accidents. There are some work to detect accidents using global positioning system (GPS) or global system for mobile (GSM) [2,3], or advanced emergency braking system (AEBS) [4] or report them via wireless [5] or cloud [6]. Taking the advantage of vehicle communications [7,8], we focus on communicating vehicles with each other and with other infrastructures to avoid accidents.

Vehicle collision avoidance system allows communication between vehicles, by alerting the driver for surrounding risks in the road. This system is very useful to road-users for multiple reasons. First, it helps in saving peoples' lives by reducing the number of crashes, injuries and deaths on the roads. Hence, it greatly increases safe driving conditions. The main contribution in this paper is to propose a vehicle collision avoidance system that can calculate the required

© Springer Nature Switzerland AG 2020
M. Belkasmi et al. (Eds.): ACOSIS 2019, CCIS 1264, pp. 53–64, 2020.
https://doi.org/10.1007/978-3-030-61143-9_5

safety distance base and can measure the nearest obstacle in the driver's way in a fog mode. The system model is implemented on small electronic vehicles, and is controlled by a driver. The proposed system is divided into two parts: the first part is the communication part, which allows drivers to share street conditions with each other. While the second part is the self-simulation, which includes measuring the safety distance base and the current distance between two vehicles in a fog mode.

2   Vehicular Communication System

Vehicular communication system is a system that is developed to allow the communication between vehicles and to provide information that helps the drivers in maintaining safety on the roads. This section contains the definition of vehicular communication system, the difference between Vehicular Ad-Hoc Network (VANET) and Mobile Ad-Hoc Network (MANET), historical background, and the advantages and disadvantages of the vehicle-to-vehicle (V2V) collision avoidance.

Vehicular Communication System is an integrated system that allows vehicles to broadcast the statuses of a road such as the speed of a vehicle, and the space between vehicles. The purpose of this system is to face critical problems such as accidents and road conditions that could happen on the road. The main aim of the system is to enable vehicles to receive warning messages, and then react immediately [9]. Using this system, vehicles can also connect to each other with a 360 degrees coverage around the car using several ways such as: Bluetooth, WLAN (Wi-Fi) and ZigBee.

The system involves two main parts, which are V2V and vehicle to-infrastructure (V2I) communications [7]. For first part of V2V, broadcasts data to the surrounding vehicles. However, the second part, V2I, is a communication method between vehicles and road infrastructure that plays a role by collecting data about traffic signals and the state of the road [10].

2.1   Design Criteria

The criteria of the proposed system consists of: usefulness, relevance, applicability, cost, parts availability, enhancement capability, and expandability. The following points define each one of them:

- Usefulness: this criterion concerns about the value of implementing the proposed system and the areas where it can provide the most benefit.
- Relevance to environment: this criterion studies if the system is relevant and needed in the environment. For example, the proposed system is not needed in a place where people do not drive or where cars are not allowed.
- Applicability: this criterion is related to the ability of applying the proposed system in the real world and if it can be easy to implement and use.
- Cost: this criterion studies the components cost.

- Parts availability: this criterion is considered in the hardware design
- Enhancement capability: this criterion considers the ability of enhancing the proposed system in the future.
- Expandability: this criterion represents the capability of changing the size of the system interaction area.

These are the criteria that guide us to implementing and evaluating the system.

2.2 V2V Standards

There are three groups of IEEE standards that work with vehicular communication system, which are IEEE 802.11p, IEEE 1609 and SAE J2735. The first type, which is IEEE 802.11p, aims to provide safety applications that help in saving people lives' and work on developing ways to solve traffic. The second type, which is IEEE 1609, includes the architecture and the management structure. The last type that is SAE J2735 which is related to the vehicular communication in which the field of this standard states the data frames and the message for the applications that uses communication. Note that in the proposed system, standards are not followed. However, we mention them as they are used in real life.

3 System Design

The proposed system consists of software and hardware parts. The hardware part involves electronic vehicles and components that are used to build the system which is shown in Fig. 1. In the software part, the components programmed by Arduino software. The system consists of Arduino [11], humidity sensor (DHT22) [12], ultrasonic sensor (SRF10) [13], Global Positioning System (GPS) module (NEO-6) [14], push buttons [15], wireless (NRF24l01) [16], buzzer and Liquid-crystal display (LCD) screen.

The system is divided into two parts: communication part and self-simulation, which are explained in details in following sections.

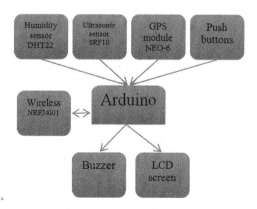

Fig. 1. System hardware components.

3.1   Communication Part

In communication part, the drivers will be able to communicate with each other by alerting each other about the surrounding hazards in the road. Three types of hazards are considered in our system model, which are: Accidents, stray animals, and puddles of water. These hazards are selected in the design because they are considered as major reason of accident-related deaths in the UAE and this can be expanded to more hazards in future developments. In this part, the proposed system has the ability to work as a transceiver for sending and receiving a message by the same system.

In the proposed transmitter design, the driver will transmit information to other drivers. The system consists of four push buttons; three push buttons are related to the three types of hazard. When the driver recognizes a hazard, he/she presses the push button that is related to that hazard. Once the push button is pressed, a warning message will be sent to the surrounding vehicles with an alerting sound. The working process of the transmitter system model is illustrated in Fig. 2. Note that, the system will wait for any of the four push buttons to be pressed, and once one is pressed, the system will immediately send the message.

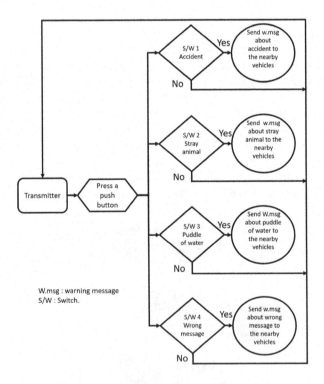

Fig. 2. Transmitter system model flowchart.

In the proposed receiver design, when a driver transmits and broadcasts an alerting signal to other drivers, the received message will be displayed on LCD screen in the form of words and letters. The receiver automatically checks if there is any available received signal, if an available signal exists, a message will be displayed on LCD screen. The proposed receiver design is displayed in Fig. 3.

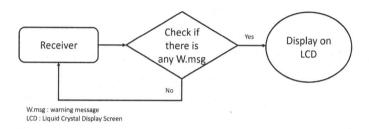

W.msg : warning message
LCD : Liquid Crystal Display Screen

Fig. 3. Receiver system model flowchart.

The communicating message that will be shared between the drivers is sent and delivered through a component called NRF24L01. The information will be transmitted from the transmitter to the receiver and it will be in form of packet. The packet format is "Enhanced Shock Burst" and it consists of address, preamble, payload, packet control, and cyclic redundancy check (CRC) field [11]. Each packet starts with the preamble, which detect the incoming packet. The preamble size is 1 byte, additionally it contains the pattern [10101010], and it used for the synchronizing purpose [11]. The address is a receiver address, and in the proposed design it has been considered that all the machines have the same address, because if one transmits all machines will receive. The packet control field consists of 6 bits payload length, 2 bits includes the packet identity which is used to know if the received packet is new or resend. The remaining 1 bit of the packet control contains the NO-ACK flag. The NO-ACK flag is used when there is an auto acknowledgment sent. The payload is the length of the load (message) and it has a size from 0 to 32 byte length. The CRC field is used to detect errors [11].

3.2   Self-simulation Part

Self- Simulation is an operation which is done by a system itself and results from the system's behavior [17]. The model executes some self-operations. These operations will be performed by the system without sharing the information to the nearest vehicles. There are two self-simulation operations that are done by the model which are measuring the nearest obstacle in fog mode and the safety distance base.

A safety distance base is the minimum space needed or required between two vehicles or between a vehicle and an obstacle. In this part, the proposed system will produce an alerting sound to the driver if he/she exceeds the safety distance base. The system is able to calculate the safety distance and to compare it with the current distance.

Measuring the safety distance requires the speed of the vehicle. The model contains a GPS module, which is called ublox NEO-6 GPS module. NEO-6 GPS module is a set of individual GPS receivers with a feature of high performance u-blox 6 engines [12]. This GPS module is programmed to measure the speed of the vehicle. After measuring the speed of the vehicle, the system will enter the speed to a safety distance equation. Table 1 which is provided by Dubai police station shows the road speed versus the required safety distance base. Based on the table readings an equation is provided in calculation and discussions part to calculate the safety distance.

A nearest obstacle is the closest obstacle in a driver's way. The proposed system measures the current distance to the nearest obstacle or vehicle only on fog mode. The nearest obstacle is necessary only in foggy conditions, because vision is not clear in this atmosphere. After measuring the distance, its value will be displayed on a screen, which is fixed close to the driver. The working process of the self-simulation system model is shown in Fig. 4.

Table 1. Road speed versus safety distance base.

Road speed (km/h)	Safety distance base (m)
60	25
80	35
100	50
120	70

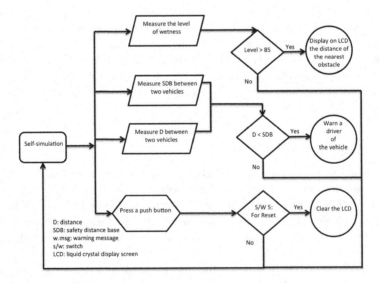

Fig. 4. Self-simulation system model flowchart.

Measuring the nearest obstacle in the fog mode will be activated via humidity sensor. The humidity sensor will be activated when the weather is foggy. The foggy condition has a certain wetness level, once this wetness level is detected, the humidity sensor activates. According to a UAE national weather website, the wetness level of the normal weather is 45%, and the wetness level of a foggy weather is 85% [18]. The measurement of the nearest obstacle will be displayed on an LCD screen, which is fixed and close to the driver.

The Schematic of the proposed system design is implemented using a Dip Trace Software. The circuit connections of the designed system that is shown in Fig. 5.

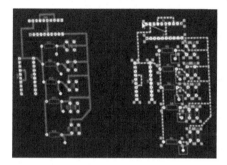

Fig. 5. The PCB connection schematic.

3.3 Calculations and Discussions

The following equation used to measure the real speed of vehicle:

$$y = 0.02066\,V - 3.002, \tag{1}$$

where y is the real spread and V represents the speed of the vehicle which is found in the car's gauge. The GPS is programmed to measure the speed of the vehicle. The real speed equation, which is shown in (1) and used in programming the system, is found through testing the GPS module.

We tested the GPS module by taking the readings of the GPS module that is connected to the system. Also, we took other readings from the speed gauge of the real car simultaneously. Then, a relationship is found by using both two readings. Consequently, the speed of the real car is approximately same as the speed that is found by using GPS with a small percentage error. Note that the accuracy of the speed of GPS module is compared by the speed which appears in the real car speed gauge. Hence, there was not huge differences, only some human error occurs when reading the speed. The relationship between a safety distance base and a speed is found by using both readings which are shown in Table 1. By using curve fitting, the first relationship between them is exponential and the second is linear, which are shown in Fig. 6, and 7.

The first equation is:

$$y = 8.8746\,e^{0.0172V}, \tag{2}$$

which is used to measure the safety distance base. The second relationship is linear which is described by:

$$y = 0.75\,V - 22.5. \tag{3}$$

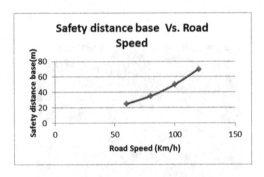

Fig. 6. Safety distance base Vs. Road speed exponential relationship.

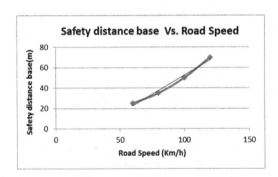

Fig. 7. Safety distance base Vs. Road speed linear relationship.

Note that in this paper, the first relationship is used in the proposed vehicle collision avoidance system, since the error percentage in (2) is 0.4%, which is less than the error percentage in (3) which is 4.5%. Note that, the error calculation is done through the percent of relative error formula, which is described by:

$$\%Error = |M - A|/A \times 100, \tag{4}$$

where M is the measured value, A is the actual value.

After measuring the safety distance, a comparison between the safety distance and the current distance will be implemented. If the current distance between two vehicles is less than the safety distance, an alarm sound will be applied. The alarm will operate continuously until the driver achieves the required safety distance. If the current distance is greater than the safety distance, it will continue to recheck the distances and make a comparison.

3.4 Economical Cost of Project Prototype

The proposed project consists of electronic components that already exist in the market and can be easily found. The total cost of the project is reasonable, because the components that are involved in the system are not very expensive. The total system will have an estimated cost of about 654 AED.

4 Results

The proposed vehicle collision avoidance system aims to spread safety in the community. The proposed system works by alerting the drivers about surrounding risks in the road, measuring the nearest obstacle in the fog mode and the safety distance base. Figure 8 displays the prototype of the proposed system.

Fig. 8. The back and front of the proposed prototype.

4.1 Results of Communication Part

Figures 9 and 10 shows the communication messages at the transmitter and the receiver, respectively.

Fig. 9. The display at the transmitter.

Fig. 10. The received message.

4.2 Results of Self-simulation Part

When the humidity sensor is activated, it measures the nearest obstacle. Figure 11 illustrates the display when the fog mode is activated. The buzzer is activated when the current distance is less than the safety distance.

Fig. 11. The display when the fog mode is activated.

4.3 Future Developments

The system can cover a larger area, by replacing the NRF24L01 component with another one that covers a wider variety of accidents causes. Furthermore, the message displayed on the LCD can be developed, in a way that the message could be in a form of picture so the non-readers could benefit from it. Also, the message that appears at the receiver could be connected with the location of the hazard so the driver can change his/her route. Moreover, if the driver exceeds the safety distance, an automatic brake will be applied instead of a manual brake. Thus, this will help the drivers that are not concentrating on the road, and it will increase the safety in a society. Note that such system already exists and it is called AEBS [4]. In addition, the system can be improved by adding advanced doors locking property in the vehicle if another vehicle is passing by. This technique allows protecting children from being impacted by the surrounding vehicles. Furthermore, detecting pedestrians and small engines such as motorcycles and bicycles can be taken into consideration for system developments.

Most of UAE accidents Road accidents are considered one of the major reasons of deaths and it is increased when the country facing foggy weather, so based on this reason, the proposed system is implemented focusing on the nearest obstacle. Therefore, the fog can be added as another hazard and the system can warn other drivers automatically because of the usage of humidity sensor. In addition, the system can be upgrade it and can consider other weather conditions to worn other drivers. Furthermore, it is not sufficient to put only one warning button since each hazards have might need different action from the driver to reduce any possible collision. However the buttons can be replaced, with voice or sign detectors. Thus, to improve the precision and response time, the system can be designed using real time architecture and programming languages. Note that, in the United States, autonomous vehicles such as Tesla use a vehicle to everything (V2X) communication system that connects any entity which affect the vehicle to a cloud and updates the traffic conditions [6]. All in all, communication protocol can be used in the future developments, so it will indicate the chain of interactions between the vehicles and the surrounding objects.

5 Conclusion

To conclude, the proposed vehicle collision avoidance system helps people in different ways. The proposed system allows the drivers to alert others about the surrounding hazards in the road. Also, it alerts the drivers if they exceeded the required safety distance base. Furthermore, the proposed system warns the drivers about a nearest obstacle in the fog mode since the vision will not be clear.

References

1. Croucher, M.: UAE has one of best road-death records in the Middle East. The National UAE (2013)
2. Sane, N.H., Patil, D.S., Thakare, S.D., Rokade, A.V.: Real time vehicle accident detection and tracking using GPS and GSM. Int. J. Recent Innov. Trends Comput. Commun. 4, 479–482 (2016)
3. Amin, M.S., Jalil, J., Reaz, M.B.I.: Accident detection and reporting system using GPS, GPRS and GSM technology. In: International Conference on Informatics, Electronics & Vision (ICIEV), pp. 640–643. IEEE, May 2012
4. United Nations: Uniform provisions concerning the approval of motor vehicles with regard to the Advanced Emergency Braking Systems (AEBS) - Addendum: 130 - Regulation: 131, 27 (2014)
5. Kannan Megalingam, R., Nammily Nair, R., Manoj Prakhya, S.: Wireless vehicular accident detection and reporting system. In: International Conference on Mechanical and Electrical Technology, pp. 636–640. IEEE (2010)
6. Zhou, H., Wang, W.: Evolutionary V2X technologies toward the internet of vehicles: challenges and opportunities, pp. 308–317. IEEE (2020)
7. Shooshtary, S.: Development of a MATLAB simulation environment for vehicle-to-vehicle and infrastructure communication based on IEEE 802.11p, Department of Technology and Built Environment, Vienna (2008)

8. Harding, J., et al.: Vehicle-to-vehicle communications: readiness of V2V technology for application (No. DOT HS 812 014). United States. National Highway Traffic Safety Administration (2014)

9. Fernandes, P., Nunes, U.: Vehicle communications: a short survey. IADIS Telecommunications, Networks and Systems, Lisboa, Portugal (2007)

10. Samad, T., Annaswamy, A.: Vehicle-to-vehicle/vehicle-to-infrastructure control from the impact of control technology. IEEE J. Mag. **31**, 26–27 (2011). www.ieeecss.org

11. Learn.sparkfun.com: Arduino Comparison Guide. https://learn.sparkfun.com/tutorials/arduino-comparison-guide#introduction

12. DHT 22 datasheet. https://www.sparkfun.com/datasheets/Sensors/Temperature/DHT22.pdf

13. Acroname.com: Devantech SRF10 Sonar Ranging Module — Acroname. https://acroname.com/products/r241-srf10?sku=R241-SRF10

14. NEO-6-ublox-datasheet. http://ecmobile.ru/user_files/File/u-blox/NEO-6_DataSheet_(GPS.G6-HW-09005).pdf

15. Arduino.cc: Arduino - Pushbutton. https://www.arduino.cc/en/Tutorial/Pushbutton

16. NRF24L01 datasheet. https://www.sparkfun.com/datasheets/Components/SMD/nRF24L01Pluss_Preliminary_Product_Specification_v1_0.pdf

17. Stellar, E., Stellar, J.: The Neurobiology of Motivation and Reward, pp. 136–137137. Springer, New York (1985). https://doi.org/10.1007/978-1-4615-8032-4

18. UAE Forecast and The National Staff: UAE weather: Humidity up to 95 percent forecast, The National. Thenational.ae (2015). http://www.thenational.ae/uae/

On the V2X Velocity Synchronization at Unsignalized Intersections: Right Hand Priority Based System

Saif Islam Bouderba$^{(\boxtimes)}$ and Najem Moussa

LAROSERI, Department of Computer Science, Faculty of Science,
University of Chouaib Doukkali, El Jadida, Morocco
{bouderba.s,moussa.n}@ucd.ac.ma

Abstract. Autonomous cars with communication capabilities allow novel techniques for controlling intersections. In recent times, cooperative intersection management is an active research subject. Both cars and intersection are capable to communicate with the aim of enhance traffic system. Several studies have exposed that the cooperative intersection management perform better than traffic lights, for the reason that the movement of cars is adapted to the current conditions and there is no need for traffic stoppage periods as with traffic lights. This work spotlights on velocity synchronization at intersections. Furthermore, a rulebased principle is made by cars approaching from various roads. When the rate injection (α) is in a low level, this permits to avoid wasting time when cars stop at intersection and therefore the velocity synchronization reduce energy consumption and increases the velocity. On the other hand, investigations exposed that certain parameters require to be studied. Thus, as the traffic density increases, cars must to be stopped and velocity synchronization is not efficient. The studied concepts are limited to First In First Served (FIFS), in this paper, the right-hand priority was extended. Novel rules are capable to adjust dynamically the entire behavior in accordance with the average velocity increase in secure conditions. Results show that the proposed approach is appropriate and efficient since it allows enhancing the average vehicle velocity while retaining a high capacity of the intersection.

Keywords: Cooperative intersection management · FIFS · V2X · Right-hand priority

1 Introduction

As an important feature in the traffic dynamics in the cities, intersections have attracted the attention of researchers. Such intersections may include traffic circle, roundabout, T-intersection, signalized intersection and unsignalized intersections, etc. [1–5]. The unsignalized intersection, in particular, has been widely studied in the literature because, First, it plays a significant role to determine

© Springer Nature Switzerland AG 2020
M. Belkasmi et al. (Eds.): ACOSIS 2019, CCIS 1264, pp. 65–72, 2020.
https://doi.org/10.1007/978-3-030-61143-9_6

road traffic capacity, particularly in rural areas. Second, car crashes are usually more recurrent at unsignalized crossroad than at crossroad with traffic lights. The rules at unsignalized crossroad have an important influence on the road traffic safety, in particular the existence of violent cars drivers. Crashes produced by violent drivers (a driver that pay no attention to the right-of-way rule) are more recurrent at unsignalized crossroad, and could cause a dangerous collision since it generally includes a right-angle impact type. The reasons of traffic collisions at unsignalized crossroads are usually complex, including human, cars mechanic and infrastructure of the roadway.

The emerging technology that installs in cars and crossroads control system can enhance the availability. In fact, the communication between cars and others infrastructures are more efficient thanks to the self-controlling cars [6,7]. A great opportunity to connect the traffic environment, the cars and the control system since we use the wireless technologies. Therefore, Cooperative Intersection Management is open up [8–11], by means of the application of novel technologies. There are remarkable novelties Within Cooperative Intersection Management compared to traditional traffic control system. In place of only city including in traditional traffic control decision, every car immediately contributes to the decision-making process. Thus, every car can be considered individually, and participate to enhance the traffic conditions according to the present observed situation, for instance, if several cars are on a road and one of them have blocked, the system will alert the other cars on the road [12]. From this opinion, each car use a simple rules individually, indeed it can participate to an entire efficient behavior at the crossroad.

Cooperative Intersection Management signified that every car exchange information with all surrounding cars to achieve a right of way. In recent times, the intersection manager (RoadSide Unit: RSU) is employed to avoid accident. In fact, it permits examining the presence list. In this situation, several approaches are offered to choose the best set of cars authorized to cross the intersection simultaneously [13,14]. Cars adjust their velocity to avoid collision with other cars approaching to the intersection. In the literature, there are several velocity synchronizations protocols. In [15,16] the time that cars occupied the potential zone of collision is reserved. Previous noticeable studies are based on acceleration communicated by the intersection (RSU) to cars that should decelerate [17,18]. In the other hand, various works depend on the presence list on which the order decides the following cars that must be take into consideration to avoid accident.

This work assumes that cars are connected and adapt their velocity according to all cars into the crossroad [19]. Furthermore, velocity synchronization is very important to avoid unnecessary speed deceleration and acceleration. However, in high traffic density velocity synchronization is not capable to perform as expected [20]. In this paper and in order to pass before than expected, cars by themselves are capable to change their ranks. The purpose of the proposed rule is to deal with various traffic density scenarios.

The remainder of this paper is structured as follows. In the following section, we provide the model definition and rule-based principle. In Sect. 3, by using our simulation program, we study the characteristics of the system where certain interesting observations are analyzed. Finally, a conclusion in Sect. 4.

2   Traffic Model

2.1   Model Structure and Rule Based Principle

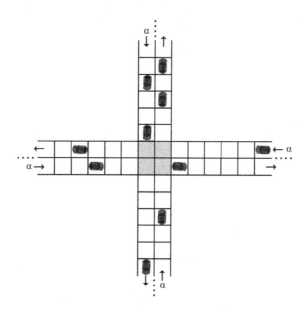

Fig. 1. Two intersecting streets, α is injection rate.

The model studied in this paper consists of two crossing roads. Both roads are two-ways, with one lane in each direction. Each road consists of L cells of identical size and the time t is discrete. All streets cross each others at the intersection sites (see Fig. 1). Open boundary conditions are applied to all roads, with four entry/exit points. We define road 1 as the road which goes from East to West, road 2 from West to East, road 3 from South to North and road 4 from North to South. Each position can be also empty or occupied by a car with the velocity $v = 0, 1, 2, 3, ..v_{max}$, where $v_{max} = 5$ is the maximum velocity. The dynamics of cars in each road is controlled by the NaSch moving rules [20] are as follows:

– Acceleration: $v = min(v + 1, v_{max})$
– Deceleration: $v = min(v, k)$ where k is the number of empty sites in front of car.
– Randomization: $v = max(v - 1, 0)$ with probability p_r.
– Advancement: the vehicle moves v sites forward.

The boundary condition is defined as follows: at each time step and for each road i, a vehicle k is injected with probability α and then the velocity $v_k = v_{max}$ is assigned to the vehicle. When vehicle k reaches the end of the driving-out lane j, it simply injected again in system. Driver behavior can be classified into three processes according to the turning probability (γ): turning to the right, left at the intersection or moving straight forward.

Algorithm 1. Turning rules for intersection

---

if (the vehicle is on the entry leg) **then**
  if ($\gamma > rand(0, 1)$) **then**
    if ($p > rand(0, 1)$) **then**
      the car will turn right
    else
      the car will turn left
    end if
  else
    the car will move straight forward
  end if
end if

2.2   Rule Based Principle

In this work, we take into account that a presence list of the other cars is associated to each car. Moreover, in the presence list, car discoveries precedent cars, their location, velocity and destination. furthermore, the cars are firstly listed in the presence list before their crosses the intersection. For the reason to dodge collision, the cars have to adjust their velocity.

2.3   Rule 1: Cooperative Drive Control

Right Hand Priority (RHP)
To manage the movement of vehicle at unsignalized crossroad, it is needed to define the priority rule when vehicle try to cross at the same time the intersection. The right-hand priority rule is the priority rule adopted by several countries, which means that the vehicles drivers in the left are obliged to yield to vehicles drivers nearing from the right at the crossroad. In this situation cars synchronize their velocity according to a presence list that respects the right-hand priority (Fig. 2).

First in First Serve (FIFS)
In this situation cars synchronize their velocity according a presence list that respects the order of the car's arrivals. In low traffic flow, cooperative velocity synchronization is very efficient. The acceleration function was enhanced [14] in order to increase the traffic.

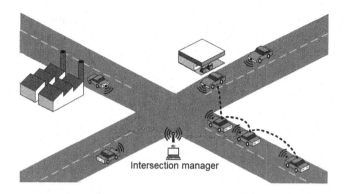

Fig. 2. Velocity synchronization at intersection: The crossroad manager regularly broadcasts the updated list according to the regularly received information from vehicle.

Either V2V communication or the intersection manager can provide information to vehicles to build their list presence. We take into consideration that each car sends regularly its position, velocity and destination to the intersection manager for safety reason and to avoid consistency problem. The car approaching to the intersection is added automatically at the lowest rank of presence list When its detected by the intersection manager. Afterwards, the intersection manager updates the data of the car, by remaining its rank.

2.4 Rule 2: Communication

Due to packet loss, delayed cars possibly will not receive the list of new obstacles. Indeed, if the car approaching the intersection has the information, it will be able to synchronize its velocity. Hence, before any change in the presence list it has to recognize the new obstacle. For safety reasons, once the achievement of the following process the car with the higher priority will be added as a new obstacle:

- According to FIFS or RHP the priority car will be added in the presence list
- According to the list received, it broadcasts informations to the first conflict cars in the presence list to request priority acknowledgement.
- If all the cars recognize the request, the classification is changed

The acknowledged status is taken into consideration by the crossroad manager to modify the rank. Furthermore, by the use of V2V communication the negotiation of the priority car for the acknowledgment can be done. In this work, we take into consideration that the manager sends request to the presence list. Hence, when the cars update their data they send the acknowledgment.

3   Results and Discussion

This section shows the simulation results for our indicators of system performance. Firstly, according to the traffic variation we take into consideration a variable traffic density, with the purpose to study the capability of RHP on the presence list. Two values of traffic densities are studied (began with $\alpha = 0.1$ up to $\alpha = 0.4$ (Fig. 3)) low traffic and (up to $\alpha = 0.8$ (Fig. 4)) high traffic. The influx changes each (200 (Fig. 4) with $\alpha + = 0.1$) (400(Fig. 3) with $\alpha + = 0.1$). In this works the curves in graphs below are achieved from the average of 50 simulation runs. Each car has three different directions to choose, we have considered ($\gamma = 0.2$) (see Algorithm 1).

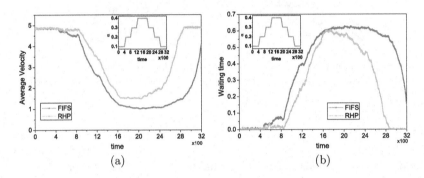

(a)   (b)

Fig. 3. The average velocity (a) and the waiting time (b) as a function of time for different methods. (Color figure online)

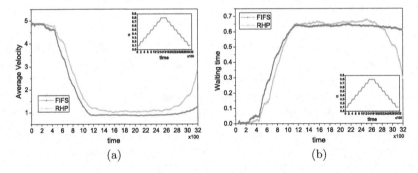

(a)   (b)

Fig. 4. The average velocity (a) and the waiting time (b) as a function of time for different methods.

Firstly, we show in Fig. 3, the average velocity as a function of the density of the system for different methods. After 400 s, one can observe from (Fig. 3) that FIFS is not able to regain its initial performance. Even if the traffic becomes smaller (after1600 s with $\alpha - = 0.1$), FIFS is not able to free the accumulated cars. RHP outperforms FIFS management approaches. The average velocity increase when the traffic density is lighter, however it remains a higher velocity even in a high traffic density (Fig. 4(a)). RHP keeps the advantage of FIFS management systems. As we can see after applied the rules, with RHP the cars make the good decision in the intersection, for that, the average velocity with the RHP (red line) give as the best result. Moreover, in Fig. 3(b), we show the waiting time as a function of the density of the system for different methods. When we increase the value of α, the density of cars increases and the road becomes crowded. This, results an increase of the waiting time of stopped cars, so we notice in Fig. 3(b) that with the RHP (red line) reduce the waiting time of cars.

4 Conclusions

A rule-based approach is presented in this work to improve the performance of cooperative intersection management. Cars can coordinate their speed to avoid wasting time when they stop at the intersection using wireless communication. Our simulation results could give us an idea about how the RHP rules proposed allows an improvement on performances more than FIFS at a high and medium density since the velocity synchronization decrease the time between cars approaching from two different lane. Both cases allowed us to Reduce energy consumption. In a future work, we aim to develop this work to a large city with several and complex interactions.

References

1. Li, X.G., Gao, Z.Y., Jia, B., Zhao, X.M.: Cellular automata model for unsignalized T-shaped intersection. Int. J. Mod. Phys. C **20**(04), 501–512 (2009)
2. Fouladvand, M.E., Sadjadi, Z., Shaebani, M.R.: Characteristics of vehicular traffic flow at a roundabout. Phys. Rev. E **70**(4), 046132 (2004)
3. Marzoug, R., Ez-Zahraouy, H., Benyoussef, A.: Simulation study of car accidents at the intersection of two roads in the mixed traffic flow. Int. J. Mod. Phys. C **26**(01), 1550007 (2015)
4. Huang, D.W.: Modeling gridlock at roundabout. Comput. Phys. Commun. **1**(189), 72–76 (2015)
5. Xie, D.F., Gao, Z.Y., Zhao, X.M., Li, K.P.: Characteristics of mixed traffic flow with non-motorized vehicles and motorized vehicles at an unsignalized intersection. Phys. A Stat. Mech. Appl. **388**(10), 2041–2050 (2009)
6. Bagloee, S.A., Tavana, M., Asadi, M., Oliver, T.: Autonomous vehicles: challenges, opportunities, and future implications for transportation policies. J. Mod. Transp. **24**(4), 284–303 (2016). https://doi.org/10.1007/s40534-016-0117-3
7. Dimitrakopoulos, G.: Current Technologies in Vehicular Communication. Springer, New York (2017). https://doi.org/10.1007/978-3-319-47244-7
8. Islam Bouderba, S., Moussa, N.: Impact of green wave traffic light system on V2V communications. In: 2018 6th International Conference on Wireless Networks and Mobile Communications (WINCOM), pp. 1–6. IEEE, 16 October 2018
9. Naranjo, J.E., García-Rosa, R., González, C., de Pedro, T., Alonso, J., Vinuesa, J.: Crossroad cooperative driving based on GPS and wireless communications. In: Moreno Díaz, R., Pichler, F., Quesada Arencibia, A. (eds.) EUROCAST 2007. LNCS, vol. 4739, pp. 1073–1080. Springer, Heidelberg (2007). https://doi.org/10.1007/978-3-540-75867-9_134

10. Bouderba, S.I., Moussa, N.: Reinforcement learning (Q-LEARNING) traffic light controller within intersection traffic system. In: Proceedings of the 4th International Conference on Big Data and Internet of Things, pp. 1–6, 23 October 2019

11. Hao, X., Abbas-Turki, A., Perronnet, F., Bouyekhf, R., El Moudni, A.: Intersection management based on V2I velocity synchronization. In: Mastorakis, N., Bulucea, A., Tsekouras, G. (eds.) Computational Problems in Science and Engineering. LNEE, vol. 343, pp. 449–470. Springer, Cham (2015). https://doi.org/10.1007/978-3-319-15765-8_28

12. Bouderba, S.I., Moussa, N.: Evolutionary dilemma game for conflict resolution at unsignalized traffic intersection. Int. J. Mod. Phys. C **30**(02n03), 1950018 (2019)

13. Ahmane, M., et al.: Modeling and controlling an isolated urban intersection based on cooperative vehicles. Transp. Res. Part C Emerg. Technol. **1**(28), 44–62 (2013)

14. Milanes, V., Villagra, J., Godoy, J., Simo, J., Pérez, J., Onieva, E.: An intelligent V2I-based traffic management system. IEEE Trans. Intell. Transp. Syst. **13**(1), 49–58 (2012)

15. Dresner, K., Stone, P.: Multiagent traffic management: a reservation-based intersection control mechanism. In: Proceedings of the Third International Joint Conference on Autonomous Agents and Multiagent Systems-Volume 2, pp. 530–537, 19 July 2004

16. de La Fortelle, A.: Analysis of reservation algorithms for cooperative planning at intersections. In: 13th International IEEE Conference on Intelligent Transportation Systems, pp. 445–449. IEEE, 19 September 2010

17. Zohdy, I.H., Rakha, H.: Game theory algorithm for intersection-based cooperative adaptive cruise control (CACC) systems. In: 2012 15th International IEEE Conference on Intelligent Transportation Systems, pp. 1097–1102. IEEE, 16 September 2012

18. Du, W., Abbas-Turki, A., Koukam, A., Galland, S., Gechter, F.: On the v2x speed synchronization at intersections: rule based system for extended virtual platooning. Procedia Comput. Sci. **1**(141), 255–562 (2018)

19. Quinlan, M., Au, T.C., Zhu, J., Stiurca, N., Stone, P.: Bringing simulation to life: a mixed reality autonomous intersection. In: 2010 IEEE/RSJ International Conference on Intelligent Robots and Systems, pp. 6083–6088. IEEE, 18 October 2010

20. Nagel, K., Schreckenberg, M.: A cellular automaton model for freeway traffic. J. Phys. I **2**(12), 2221–2229 (1992)

Convolutional Neural Networks for Traffic Signs Recognition

Btissam Bousarhane$^{(\boxtimes)}$ (iD) and Driss Bouzidi (iD)

Smart Systems Laboratory (SSL), National School of Computer Science and
Systems Analysis ENSIAS, Mohammed V University, Rabat, Morocco
{ibtissam_bousarhane,driss.bouzidi}@um5.ac.ma

Abstract. The application fields of traffic signs recognition are multiple, including autonomous vehicles, self-driving cars, Advanced Driver Assistance Systems (ADAS), etc. The ultimate goal of these systems is to save lives on roads by reducing the number of accidents. Specifically, the incidents caused by drivers' distraction and inattention. Efficient traffic signs recognition systems represent then one of the major components in making cars safer, not only for drivers, but for pedestrians as well. In this domain of research, Deep Learning approaches have gained more popularity due to their appealing performances. However, these approaches still face many difficulties, especially, those related to adverse conditions and real time response. Furthermore, these methods are very demanding in terms of hardware requirements, computational load and training data. From this perspective, our work aims to create an efficient Deep Learning model, for the classification of traffic signs, under adverse and challenging conditions. The objective of the research is to reach a high recognition accuracy, via a faster learning process, while using limited data and hardware resources. The obtained results show that CNNs could reach an accuracy of more than 99.6% for signs with no challenges. However, their accuracy decreases when dealing with more challenging traffic signs. The results shows also that lighter architectures are efficient in terms of generalization, accuracy and speed.

Keywords: Traffic signs recognition · Classification · Deep learning · CNN · Adverse conditions

1 Introduction

Objects recognition is a complex task for which many approaches are adopted by researchers. This recognition includes pedestrians, cars, lanes, traffic signs, etc. In terms of recognition performances, recently Deep Learning methods have proven their superiority, especially CNNs [1].

Despite of their high accuracy, that outperforms in some cases human performances, their use for recognizing objects, under adverse conditions, still presents a challenge, for which researchers are still looking for answers. To face this problem, using a huge dataset with thousands of images is a solution, which some researchers opt for [2]. This option could, surely, improve the recognition performances. However, it is very

© Springer Nature Switzerland AG 2020
M. Belkasmi et al. (Eds.): ACOSIS 2019, CCIS 1264, pp. 73–91, 2020.
https://doi.org/10.1007/978-3-030-61143-9_7

computationally expensive and tightly related to the hardware performances and optimization [3]. Thing which presents, on the one hand, a real issue for researchers, who aim to benefit from the outstanding results of these approaches in solving complex problems. On the other hand, this challenge will also limit their use in low resources systems, such as mobile devices, etc.

The hardware requirements represent, in fact, just one of the challenges that still face researchers in this domain. Another faced problem is the real time recognition of Deep Learning algorithms, because these approaches are very time consuming, in comparison to other methods. In addition to the response time in real world situations, the speed of the training process is another faced issue. Indeed, we should wait for a long period of time until the accomplishment of the process. However, if the model is not performant enough, we should repeat the process, again and again, until reaching satisfactory results. By consequence, the operation of validation could take weeks and even months, which makes the task for researchers more and more complicated.

Moreover, collecting a huge number of images represents also a hard task, in addition to the process of annotation that makes the operation even more and more difficult. From that, an important question raises about whether Deep Learning approaches need or not a huge amount of data to reach high accuracy? And whether the type of training data and its structure present, instead, the key element for training a high performant model?

From this perspective, and knowing that the recognition process includes two principal stages, which are detection and classification. Our research aims to train an efficient model for traffic sign classification, and that by taking into consideration, on the one hand, a reasonable amount of data, and by adopting on the other hand, a faster learning and inference process. Hence, the rest of the paper is organized as follows: Sect. 2 presents some related works. The proposed approach for traffic signs classification and the datasets used during the training process are presented in Sect. 3. Section 4 presents the testing datasets and the obtained results. Section 5 illustrates the impact of training data on classification accuracy. Finally, Sect. 6 includes concluding remarks and future works.

2   Related Works

Traffic signs recognition is widely treated in the literature. Thus, in this section we will, briefly, present some works related to this subject. A more detailed literature review is presented in the works of Safat et al. [4], Gokul et al. [5], Sumi and Arun [6], Bousarhane, Bensiali and Bouzidi [7], etc.

Generally, traffic signs recognition approaches can be grouped into three main categories, which are detection approaches, classification approaches, and combined approaches.

2.1   Detection Approaches

For this type of approaches, they focus essentially on locating traffic signs on real scene images. This detection represents, in fact, the most difficult stage within the recognition process. This difficulty is mainly related to the complexity of the road scene, and also to the changes that affect traffic signs shapes and colors. These changes are caused, indeed, by several external factors, as for example: illumination variations, weather conditions, angle of vision, distance, noise, exposure, etc.

In this same context, we can mention the work of Liu, Chang and Chen [8], which aims to optimize the traffic signs detection through the reduction of false positives. Their approach consists of extracting HOG regions of interest (Histogram of Oriented Gradient) and MBLBP (Multi-radius Block Local Binary Pattern) descriptors. The genetic algorithm is also used with SVM (Support Vector Machine) to improve the performances. The obtained results show that the method can reach 99.2% in terms of accuracy (the test is performed using a Chinese database).

For Singh, Solanki and Dixi [9], another descriptor is used, which is more precisely SURF (Speeded-Up Robust Features). To improve the obtained results and reduce the noise, they have also used the RANSAC algorithm (Random Sample Consensus). According to the authors, the adopted approach is robust, especially, when it comes to rotations and occlusions problems.

Concerning the real-time detection, we can mention the work of Boumediene et al. [10], which proposes an integrated system for traffic signs detection. The proposed approach is based on signs tracking in the next frames. For the authors, the presented method reduces, significantly, the rate of false positives.

Adopting almost the same approach, we find the work of Lafuente-Arroyo et al. [11], which use also a prediction filter for traffic signs positions. The method is based on two video sequences, which have been taken under adverse conditions. The goal of the method is to reduce also false alarms 'rate'. The obtained results show that false alarms number, tracked by the algorithm, is very small in both sequences.

Another performant method used in this stage is based on Neural Networks, and more specifically CNNs, as presented in the work of Zhang et al. [12]. Inspired by YOLOv2, the work presents a real time detection algorithm for Chinese traffic signs, using a Deep Convolutional Network. The approach is based on the creation of finer features maps, in order to efficiently detect small traffic signs. The evaluation of the approach is realized using Chinese Traffic Sign Dataset (CTSD) and German Traffic Sign Detection Benchmark (GTSDB). The obtained results show that the proposed method is fast and robust, where each image is detected within only 0.017 s.

2.2 Classification Approaches

For this type of approaches, they are rather interested in the identification of the information within the road signs. To realize this identification or classification, these methods are, generally, based on images that already include detected traffic signs.

In this context, we can mention, for example, the work of Huang et al. [13]. The authors propose a system for traffic signs classification, based on Extreme Learning Machine (ELM) algorithm. The evaluation of this method has been realized on three databases, which are more exactly German TSRB, Belgium TSC and MASTIF Revised. The algorithm reaches almost 99% as recognition rate on GTSRB, and more than 98% on BTSC dataset.

Concerning CNNs, they are used in the classification stage, as well as in the detection one, as presented in many researches [14]. Among these works, we find for example, the work of Mehta, Paunwala and Vaidya [15], which propose a Deep approach for the classification of traffic signs, based on CNNs. The evaluation of the proposed approach is realized using Belgium Traffic Sign Dataset (BTSD). The method achieves competitive results, using Adam Optimizer (Adaptive Moment Estimation) and Softmax activation function.

2.3   Recognition Approaches

This type of methods focuses, instead, on the two stages: detection and classification.

Indeed, many works adopt this type of approaches [16]. Among these researches, we can mention the work of Ursic et al. [17]. The authors propose a method that is based on Faster R-CNN (Faster Region-based Convolutional Neural Network). The authors aim to ensure the detection and the classification of more than 100 categories of traffic signs (integrated in a new dataset of images acquired from Slovenian roads). The obtained results reach 90% in terms of precision.

For Ramadhan and Ergen [18], the authors propose another method for the detection and the classification of traffic signs. The method is based on Blob analysis algorithm. For the detection stage, it is based on the shapes and colors of extracted regions of interest. For the classification stage, it is realised by comparing the extracted regions with a database that contains almost 600 images. The detection rate obtained is 94.95%, while the classification rate is 98%.

Ellahyani, El Ansari and El Jaafari [19] have adopted a different method that is based on the segmentation in the HSV color space. The authors have used also Distance to border feature and Random Forests classifier to detect triangular, circular and rectangular shapes in the segmented images. For the identification of the information within the signs, the authors have compared four descriptors: HOG, Gabor, LBP and LSS. For the classification, they have realized also a comparison between Random Forests and Support Vector Machine. According to the authors, the best results are obtained by combining HOG with LSS and Random Forests classifier. The method is tested on the Swedish Traffic Signs Dataset, and the classification rate obtained exceeds 96%.

In the light of the realized review, we find that there are multiple methods for traffic signs recognition. These methods can be grouped into two principal approaches, which are: classical and machine learning approaches.

For the classical ones, they are essentially based on traffic signs colors and shapes. However, this type of approaches faces many difficulties. In effect, color based approaches are very sensitive to illumination changes, weather conditions, fading colors, etc. While we find that shape based approaches are, on the contrary, very sensitive to occlusions and damaged signs.

Given these points, Machine Learning approaches are adopted, instead, as a solution to these challenges. However, in spite of their high performances, they need in contrast many classifiers to ensure this recognition. In addition to that, their performances depend tightly on the quality of extracted features [19].

Where comes the need for adopting more performant methods, as Deep Learning approaches to deal with all these difficulties. Indeed, the literature review shows that Deep Learning approaches are more efficient and accurate, in comparison to existing methods. Moreover, they are also more suitable for dealing with multiclass classification problems, especially CNNs [7].

For these specific reasons, we have opted for Convolutional Neural Networks to realize our classification model.

3   Adopted Model and Training Datasets

In many fields of research, Deep Learning approaches have proven their superiority for classification problems, as presented in the works of Aceto et al. [20, 21] and [22], especially Convolutional Neural Networks.

While CNNs are widely used by many researchers, we find that there is no standard way to choose CNNs depth and size. Hence, for our approach, the choice of these parameters will be, essentially, guided by the objective of our work, which consists on ensuring the efficiency of the training and the inference stages. By efficiency, we mean more precisely, the generalization, the speed and the accuracy of the model.

Generally, a Neural Network is composed of input, hidden and output layers. For our approach, we have used a multi-layer Neural Network, because it is more suitable for problems that are linearly inseparable, like images classification.

Concerning the input images, using the original ones leads generally to low results. In contrast, manually extracting features from the images is a hard and time consuming task. Furthermore, the obtained results depend tightly on the type of the used features. For these specific reasons, we have opted for an automatic extraction of features using, instead, a Convolutional Neural Network.

CNNs include features extractor, which represents in fact their main advantage, in comparison to other types of techniques. Generally, more deeper convolutional layers could lead to better precision results. However, taking into consideration that we aim to get a faster learning & inference process (using low resources hardware), we have opted just for two convolutional layers.

The goal of the two layers is to generate a certain number of feature maps, which accentuate the unique features of the original images. The values of the convolutional filters are determined via the training process.

In order to prevent overfitting, and reduce the computational load, two pooling layers are added. The role of the pooling layers is to reduce the dimension of the images, and that by combining neighboring pixels of the images to a more representative value. The pooling pixels are selected from the square matrix, and their value is set as the maximum of the selected pixels.

For the hidden layers, we have opted for two layers to reduce the computational time and load, and also to prevent overfitting problems.

Concerning the training process, it consists on modifying the model in order to reduce the error between its output and the correct output. This process is realized by continually changing the weights of the Neural Network. To ensure that, a cost function should be used to measure the error of the network.

For our approach, we have used Cross Entropy loss [23] and [24] to face the vanishing gradient problems. Indeed, this function is more sensitive to the error, and provides higher performances during the training process. Furthermore, it reduces the training error at a much faster rate, and yields hence a faster learning process.

In addition to that, the Rectified Linear Unit (ReLU) function is used as the activation function [25] and [26], because it insures better transmission of the error (Fig. 1).

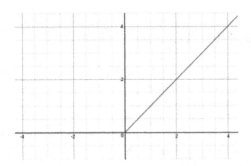

Fig. 1. The Rectified Linear Unit (ReLU) activation function

$$f(x) = \begin{cases} 0 & for\, x < 0 \\ x & for\, x \geq 0 \end{cases} \tag{1}$$

$$f'(W) = -\sum_{i=1}^{m} \tilde{y}_i \log(y_i) \tag{2}$$

W: Training weight matrix
m: Total number of classes
y_i: i-th prediction class
$\tilde{y}_i$:   i-th true class of training samples

For error calculations, we have used mini Batch [27], because it combines between the speed of the Stochastic Gradient Descent (SGD), and the stability of the Batch method. We have used also a dropout [28] technique to train only randomly selected neurons from the hidden layers. That will contribute to face overfitting problems, and will also reduce the required time for the training process.

For the final layer, the number of output nodes matches the number of classes. Furthermore, the Softmax [29] function is used as the activation function for this layer, because it generally leads to better results.

In this work, we have handled two types of classification problems: a binary classification (2 traffic signs classes), and a multiclass classification (for five classes of road signs) (Fig. 2).

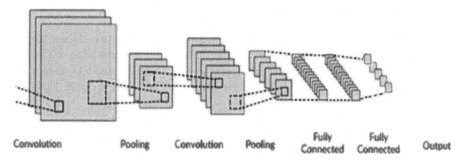

Fig. 2. Convolutional Neural Network for multiclass classification

As mentioned in the literature review [7], there are many public datasets for traffic signs recognition. However, the majority of these datasets don't reflect, exactly, the real world conditions faced by the recognition process. To overcome these limitations, we will explore a new dataset, which is more exactly "The Challenging Unreal and Real Environments for Traffic Sign Recognition" [30]. This dataset includes five levels of challenging conditions, that vary from mild to severe: rain, snow, haze, brightness, darkness, shadow, blur, decolorization, dirty lens, noise, etc.

In CURE-TSR each challenge has a number between 1 & 5, where 1 is the least severe conditions and 5 is the most severe challenges. For our case, we will explore three levels of the dataset, which are level 1, level 2 and signs with no challenges. The figure below shows the different levels of CURE-TSR dataset (Fig. 3).

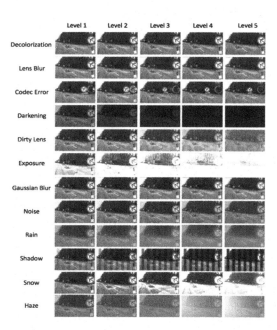

Fig. 3. Challenges types and levels in CURE-TSR. Source: https://ghassanalregibdotcom. wordpress.com/cure-tsr/

3.1   Dataset for Training 2 Classes

For the binary classification, we have used a dataset that contains two classes for the training, which are No Left and No Entry signs. This dataset is extracted from CURE-TSR, and it contains 1 893 images: 950 images for No Left signs and 943 images for No Entry signs, as shown in the table and the figures below (Figs. 4, 5 and Table 1).

No Left signs   No Entry signs

Fig. 4. Signs used for the binary classification

Table 1. Training dataset for 2 classes

Signs types	Images	Total
No Left signs	950	1 893
No Entry signs	943	

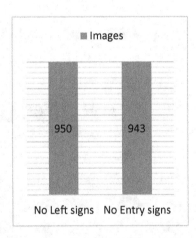

Fig. 5. Balance between the 2 training classes

We have trained our model using this dataset. The figure below shows the predictions of the model against the ground truth (Fig. 6):

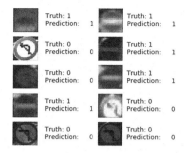

Truth: 1		Truth: 1	
Prediction:	1	Prediction:	1
Truth: 0		Truth: 1	
Prediction:	0	Prediction:	1
Truth: 0		Truth: 1	
Prediction:	0	Prediction:	1
Truth: 1		Truth: 0	
Prediction:	1	Prediction:	0
Truth: 0		Truth: 0	
Prediction:	0	Prediction:	0

Fig. 6. Predictions of the model against the ground truth (2 classes)

3.2 Dataset for Training 5 Classes

For the multiclass classification, we have used a dataset that includes five classes, extracted also from CURE-TSR. These classes are more precisely: No Overtaking, No Stopping, No Parking, No Entry and Stop signs. The total number of training images is 969, as shown in the figure below (Fig. 7).

Fig. 7. Signs used for the multiclass classification

The number of images in each class is mentioned in the table and the figure below (Table 2 and Fig. 8).

Table 2. Training dataset for 5 classes

	Signs types					
	No Overtaking	No Stopping	No Parking	Stop	No Entry	Total
Number of training images	150	191	264	170	194	969

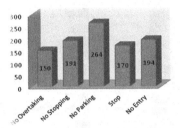

Fig. 8. Balance between the 5 training classes

The predictions of the model against the ground truth are obtained after the training of our model, as the figure below shows (Fig. 9).

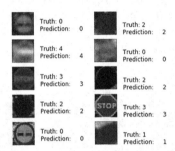

Fig. 9. Predictions of the model against the ground truth (5 classes)

4   Testing Datasets and Obtained Results

To evaluate the performances of our model, we have used multiple datasets (CURE-TSR level 1, 2 & sings with no challenges).

4.1   Dataset for Testing the 2 Classes Model

To test the model, we have used 897 images randomly extracted from the used dataset. 619 for No Left signs and 278 for No Entry traffic signs. The obtained results overpass 99% in terms of accuracy, as mentioned in the table below (Table 3).

Table 3. Testing dataset for the 2 classes model

	Signs type		
	No Left	No Entry	Total
Number of images	619	278	897
Accuracy	**99.66%**		

To prove the effectiveness of our method, we have used a more challenging images for testing the presented approach. This dataset includes more than 8 800 images, 6 190 for No Left signs and 2 631 for No Entry signs. The overall accuracy obtained is 98.18%, as presented in the table below (Table 4).

Table 4. Second dataset for testing the 2 classes model

	No Left	No Entry	Total
	6 190	2 631	8 821
Accuracy	98.18%		

To further evaluate the performances of the presented approach against many adverse conditions, we have added to the dataset above more than 2 000 images of traffic signs that contain more difficulties and challenges for the classification process. These challenges are more precisely: shadow, exposure, snow, darkening, decolorization, Gaussian blur, lens blur, noise, dirty lens, rain, codec error and haze.

The total number of this testing dataset is 10 572 images, and the number of traffic signs representing each adverse condition, is presented in the table below (Table 5).

Table 5. Number of images representing each adverse condition

	Adverse conditions	Traffic signs type	
		No Left	No Entry
Images number	Shadow	619	278
	Exposure	619	278
	Snow	619	278
	Darkening	619	278
	Decolorization	619	278
	Gaussian blur	619	278
	Lens blur	619	278
	Noise	619	278
	Dirty lens	619	278
	Rain	619	321
	Codec error	619	278
	Haze	619	43
	Total	**7 428**	**3 144**
		10 572	

The figure below shows some of the signs, with challenging conditions, extracted from the testing dataset (Fig. 10).

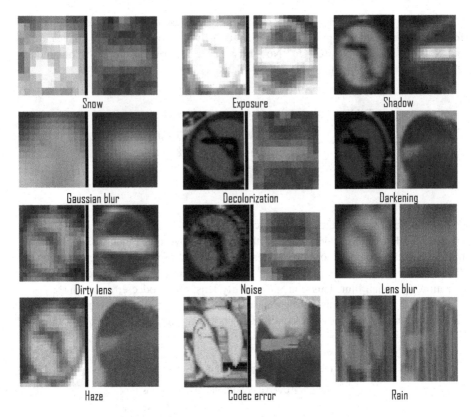

Fig. 10. Challenging conditions in the testing dataset

The table below presents the accuracy obtained for the classification of each of the adverse conditions mentioned above (Table 6).

Table 6. Accuracy obtained for each adverse condition

Signs with adverse conditions	Accuracy
Shadow	99,66%
Exposure	99,66%
Snow	99,66%
Darkening	99,66%
Decolorization	99,66%
Gaussian blur	99,66%
Lens blur	99,66%
Noise	99,66%
Dirty lens	99,66%
Haze	99,39%
Rain	90,63
Codec error	87,4%
	97,86%

From this table, we find that traffic signs affected by rain and codec errors present more difficulties for the classification model, with 90% and 87% respectively. On the other hand, we find that the other signs are classified with high accuracy, which exceeds 99%, as shown in the figure below (Fig. 11).

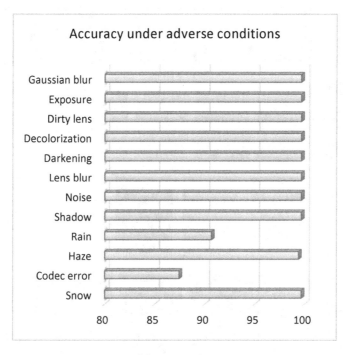

Fig. 11. Accuracies obtained in challenging conditions

4.2 Dataset for Testing the 5 Classes Model

For the five classes classification, we have used a dataset that includes 3 769 images. The figure below presents the number of traffic signs for each of these five classes: No Overtaking, No Stopping, No Parking, Stop, and No Entry (Fig. 12).

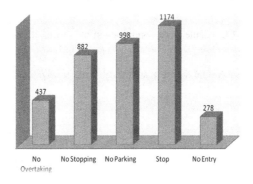

Fig. 12. Testing dataset for the 5 classes model

The figure below shows the type of adverse conditions in this testing dataset (Fig. 13).

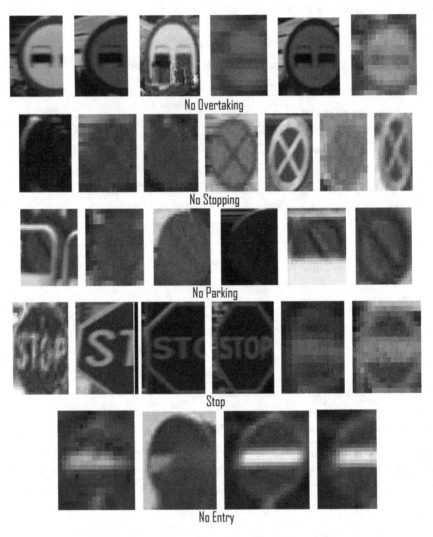

Fig. 13. Challenging conditions in the testing dataset

The approach has obtained a high accuracy with 97.87% as shown in the table below (Table 7).

Table 7. Accuracy obtained for the five class model

	Signs types					
	No Overtaking	No Stopping	No Parking	Stop	No Entry	Total
Number of training images	437	882	998	1 174	278	3 769
Accuracy	**97.87%**					

As already presented, the existing public datasets present many limitations, and the majority of these datasets don't really reflect the adverse conditions, faced in real world scenarios [30] and [31]. While we find, on the contrary, that CURE-TSR includes more challenging conditions.

According to that, to compare our approach with other existing ones, the comparison should be realized with methods that use also CURE-TSR. However, this dataset is a new one, and it is not yet widely used by researchers. Hence, Table 8 shows a comparison between the adopted approach and some other approaches that use different public datasets. This comparison includes the works that are, particularly, designed for real time classification of traffic signs. For state of the art methods, like AlexNet [32], VGG16 [33], etc. they are also very performant. However, these architectures are very computationally expensive, and the parameters used, in these approaches, exceed millions of parameters (60 M and 138 M, respectively).

Table 8. Comparison between our approach and some real time classification approaches

Reference	Time	Number of CNNs	Training images	Accuracy	Configuration
(Chen, et al. 2017)	2.7 ms	2 CNNS	26 640 (GTSRB)	95.22%	3.4 GHz CPU
			9 times GTSRB (266 400)	98.26%	
Our approach	400 us	1 CNN	969 (CURE-TSR)	97.87%	2.0 GHz CPU
(Yang et al. 2016)	3 ms	3 CNNS	50 times GTSRB	97.75%	3.7 GHz CPU, 4 Core
(Vennelakanti et al. 2019)	–	1 CNN	10 582 (BTSRD+GTSRB)	95.30%	GPU
		3 CNNS		98.11%	

Table 8 shows that each sign is recognized within 400 µs in our approach, within 3 ms in the approach of Yang et al. [16], and within 2.7 ms in the approach of Chen et al. [34].

We find also that Chen et al. and Yang et al. [16] have used multiple CNNs instead of one. To further improve their performances, they have used also a huge number of training data (9 and 50 times data) from German Traffic Signs Recognition Benchmark (GTSRB).

Therefore, after data augmentation they have got respectively 98.26% and 97.75% as classification accuracy.

For Vennelakanti et al. [35], they have opted for one CNN in their first approach. They have used also a big number of training data. Concerning the obtained results, their approach gets an accuracy of 95.30% using GPU. To improve the performances (in order to reach 98.11%), they have adopted a second approach based, instead, on 3 CNNs.

5   Impact of Training Data on Classification Accuracy

To know if there is a correlation between the training data and the classification accuracy, we have trained the five classes' model multiple times, and that by changing every time the number of training data. For the testing process, we have used the same number of testing data (3769) each time.

As mentioned in the previous section, the accuracy obtained for training 969 images is 97.87%. In fact, we have obtained a higher accuracy with 98.16%, by using less (873) images for training the same previous model. Furthermore, when training the model with even more less images (700), we have got the same accuracy obtained when using 873 images, with 98.16% (which represents the higher accuracy obtained by the adopted model).

These results show us that, using more training data don't necessarily improve the classification accuracy. In contrast, we can use less images and obtain higher accuracies. In other words, we can say that for a CNN, the structure and the diversity of the training samples is, instead, much more important than the number of images in the training datasets.

On the other hand, we find that using a light CNN architecture could outperform deeper ones for solving some classification problems. Knowing that, in the adopted approach we have used a small number of kernels and nodes for convolutional and fully connected layers. Thing that makes the training process much faster, and the inference speed does not exceed 400 µs per image (using CPU, 2.0 GHz).

6   Conclusion

The objective of our work is to train a high performant model for traffic signs classification, using a reasonable amount of data, in order to insure a faster learning process.

The obtained results show us that CNNs are very efficient for both binary and multiclass classification problems. They have proven also their high performances in the classification of traffic signs with no challenges, and also for traffic signs affected by adverse conditions.

The results show also that, there is no need for a very deep model, with many layers and thousands of nodes, to get high performances. Instead, using a compact model with a very structured dataset is enough to get satisfactory results (that exceed generally 97%). Thing which helps to reduce, enormously, the training time, and also the hardware requirements for training and testing CNN models.

To evaluate our approach, we have used just a part of CURE-TSR dataset. For that, we plan to explore the whole testing dataset for our next work. Furthermore, in order to ameliorate the obtained results we will add, during the training stage, more challenging images from CURE-TSR dataset. Especially level 5 images, which contain more severe challenges for the classification. We will add also more types of traffic signs, to include the totality of the categories presented in this dataset, which are more precisely 14.

On the other hand, and knowing that CURE-TSR is a new dataset, we will use also other existing public datasets, that are widely used by researchers to prove the effectiveness of the adopted approach.

References

1. Stallkamp, J., Schlipsing, M., Salmen, J. Igel, C.: Man vs. computer: benchmarking machine learning algorithms for traffic sign recognition. Neural Netw. **32**, 323–332 (2012)
2. Wong, S.C., Gatt, A., Stamatescu, V. McDonnel, M.D.: Understanding data augmentation for classification: when to warp? In: International Conference on Digital Image Computing: Techniques and Applications (DICTA), Gold Coast, QLD, Australia (2016)
3. Li, J., Zhang, C., Cao, Q., Qi, C., Huang, J., Xie, C.: An experimental study on deep learning based on different hardware configurations. In: International Conference on Networking, Architecture, and Storage (NAS), Shenzhen, China (2017)
4. Safat, B., Mohammad, A., Aini, H., Salina, A.: Comparative survey on traffic sign detection and recognition: a review. Przegląd Elektrotechniczny, R. 91(NR 12/2015) (2015)
5. Gokul, S., Suresh Kumar, S., Giriprasad, S.: Survey - an exploration of various techniques for sign detection in traffic panels. ARPN J. Eng. Appl. Sci. **10**(9) (2015)
6. Sumi, K., Arun, K.: Detection and recognition of road traffic signs: a survey. Int. J. Comput. Appl. **160**(3) (2017)
7. Bousarhane, B., Bensiali, S., Bouzidi, D.: Road signs recognition: state-of-the-art and perspectives. Int. J. Data Anal. Tech. Strategies. Spec. Issue. Adv. Appl. Optim. Learn. (in press). https://www.inderscience.com/info/ingeneral/forthcoming.php?jcode=ijdats
8. Liu, C., Chang, F., Chen, Z.: Traffic sign detection based on regions of interest and HOG-MBLBP features. J. Electron. Inf. Technol. **38**(5) (2016)
9. Singh Solanki, D., Dixit, G.: Traffic sign detection using feature based method. Int. J. Adv. Res. Comput. Sci. Softw. Eng. **5**(2) (2015)
10. Boumediene, M., et al.: Coupled detection, association and tracking for traffic sign recognition. In: IEEE, Intelligent Vehicles Symposium Proceedings (2014)
11. Lafuente-Arroyo, S., et al.: Road sign tracking with a predictive filter solution. In: 32nd IEEE Annual Conference on Industrial Electronics, IECON 2006 (2006)
12. Zhang, J., Huang, M., Jin, X., Li, X.: A real-time chinese traffic sign detection algorithm based on modified YOLOv2. Algorithms **10**(4), 127 (2017)
13. Huang, Z., et al.: An efficient method for traffic sign recognition based on extreme learning machine. IEEE Trans. Cybern. **47**(4), 920–933 (2017)
14. Stallkamp, J., Schlipsing, M., Salmen, J., Igel, C.: The German traffic sign recognition benchmark: a multi-class classification competition. In: The International Joint Conference on Neural Networks (IJCNN). IEEE (2011)
15. Mehta, S., Paunwala, C., Vaidya, B.: CNN based traffic sign classification using adam optimizer. In: International Conference on Intelligent Computing and Control Systems (ICCS), Madurai, India. IEEE (2019)

16. Yang, Y., Luo, H., Xu, H., Wu, F.: Towards real-time traffic sign detection and classification. IEEE Trans. Intell. Transp. Syst. **17**(7), 2022–2031 (2016)
17. Ursic, P., et al.: Towards large-scale traffic sign detection and recognition. In: 22nd Computer Vision Winter Workshop, Austria (2017)
18. Ramadhan, S.O., Ergen, B.: Traffic sign detection and recognition. Int. J. Adv. Res. Electr., Electron. Instrum. Eng. **6**(2) (2017)
19. Ellahyani, A., EL Ansari, M., El Jaafari, I.: Traffic sign detection and recognition using features combination and random forests. Int. J. Adv. Comput. Sci. Appl. **7**(1) (2016)
20. Aceto, G., Ciuonzo, D., Montieri, A., Pescapé, A.: Mobile encrypted traffic classification using deep learning. In: IEEE, Network Traffic Measurement and Analysis Conference (TMA) (2018)
21. Aceto, G., Ciuonzo, D., Montieri, A., Pescapé, A.: Mobile encrypted traffic classification using deep learning: Experimental evaluation, lessons learned, and challenges. IEEE Trans. Netw. Serv. Manag. **16**, 445–458 (2019)
22. Aceto, G., Ciuonzo, D., Montieri, A., Persico, V., Pescapé, A.: Know your big data trade-offs when classifying encrypted mobile traffic with deep learning. In: IEEE, Network Traffic Measurement and Analysis Conference (TMA) (2019)
23. Tong, W., Pi-Lian, H.: The classification algorithm based on the cross entropy rule and new activation function in fuzzy neural network. In: International Conference on Machine Learning and Cybernetics, Guangzhou, China (2005)
24. Zhou, Y., Wang, X., Zhang, M., Zhu, J., Zheng, R., Wu, Q.: MPCE: a maximum probability based cross entropy loss function for neural network classification. IEEE Access, **7**, 146331–146341 (2019)
25. Hara, K., Saito, D., Shouno, H.: Analysis of function of rectified linear unit used in deep learning, In: International Joint Conference on Neural Networks (IJCNN), Killarney, Ireland (2015)
26. Nwankpa, C., Ijomah, W., Gachagan, A., Marshall, S.: Activation functions: comparison of trends in practice and research for deep learning (2018)
27. Ito, D., Okamoto, T., Koakutsu, S.: A learning algorithm with a gradient normalization and a learning rate adaptation for the mini-batch type learning. In: IEEE, 56th Annual Conference of the Society of Instrument and Control Engineers of Japan (SICE), Kanazawa, Japan (2017)
28. Shen, J., Shafiq, O.: Deep learning convolutional neural networks with dropout - a parallel approach. In: 17th IEEE International Conference on Machine Learning and Applications (ICMLA), Orlando, FL, USA (2018)
29. Kouretas, I., Paliouras, V.: Simplified hardware implementation of the softmax activation function. In: IEEE, 8th International Conference on Modern Circuits and Systems Technologies (MOCAST), Thessaloniki, Greece (2019)
30. Temel, D., Kwon, G., Prabhushankar, M., AlRegib, G.: CURE-TSR: challenging unreal and real environments for traffic sign recognition. In: 31st Conference on Neural Information Processing Systems (NIPS), Machine Learning for Intelligent Transportation Systems Workshop, Long Beach, CA, USA (2017)
31. Zhu, Z., Liang, D., Zhang, S., Huang, X., Li, B., Hu, S.: Traffic-sign detection and classification in the wild. In: IEEE, Computer Vision and Pattern Recognition (CVPR) (2016)
32. Krizhevsky, A., Sutskever, I., Hinton, G.: ImageNet classification with deep convolutional neural networks. In: Advances in Neural Information Processing Systems (NIPS) (2012)
33. Simonyan, K., Zisserman, A.: Very deep convolutional networks for large-scale image recognition. In: 3th International Conference on Learning Representations (ICLR) (2015)

34. Chen, L., Guanghui, Z., Junwei, Z., Li, K.: Real-time traffic sign classification using combined convolutional neural networks. In: IEEE, 4th IAPR Asian Conference on Pattern Recognition (ACPR), Nanjing, China (2017)
35. Vennelakanti, A., Shreya, S., Rajendran, R., Sarkar, D., Muddegowda, D., Hanagal, P.: Traffic sign detection and recognition using a CNN ensemble. In: EEE, International Conference on Consumer Electronics (ICCE), Las Vegas, NV, USA (2019)

Channel Coding

An Efficient Semi-Algebraic Decoding Algorithm for Golay Code

Hamza Boualame[1]([envelope]) [ID], Idriss Chana[2], and Mostafa Belkasmi[1] [ID]

[1] ICES Team, ENSIAS, Mohammed V University in Rabat, Rabat, Morocco
boualame.hamza@gmail.com, mostafa.belksami@um5.ac.ma
[2] LMMI Laboratory, ENSAM, Research Team: ISIC ESTM,
Moulay-Ismail University, Meknes, Morocco
idrisschana@gmail.com

Abstract. In this paper, we propose an algorithm for decoding the QR(23,12,7) code knowing by the binary Golay code. This method never calculates unknown syndromes and does not need the error locator polynomial. Indeed, we use simple parameters to locate the error position. So, to validate the proposed method, all possible error patterns are tested and the proposed decoder correct all of them.

Keywords: QR codes · Golay code · Hard decoding · Syndrome decoding

1 Introduction

First and foremost, we want to signalize that this work is a continuation of that published in [1]. It's a second version when we propose a modified non-algebraic decoding algorithm for binary quadratic residue QR code. Then, the QR codes, introduced by Prange in 1958 [2], belong to the family of cyclic codes. They have a code rate $R \geq 1/2$ and generally possess very good error correction capabilities. They are the powerful codes known in coding theory, due to their particular mathematical structure, which give a good resistance against the errors produced by the channel. For this reason, such codes can be applied to a very noisy channel [3]. However, the QR codes present a complex algebraic structure, which makes them difficult to decode [4].

The difficulty resides in the way that each QR code has specific sets of conditions for determining the errors positions. In fact, a notable work has been made in the last years, in which, the most-known using decoding techniques can be divided into two general categories, namely algebraic and non-algebraic techniques. The former, based on the fact that the error-locator polynomial $L(z)$ can be found if the needed elimination processes to solve a nonlinear equations are carried out by using Sylvester's resultant [5] or Gröbner basis [6]. Then, substituting $L(z)$ by the roots of the generator polynomial leads to locate the errors positions [7,8]. Though, this process is certainly not obvious as we don't

© Springer Nature Switzerland AG 2020
M. Belkasmi et al. (Eds.): ACOSIS 2019, CCIS 1264, pp. 95–102, 2020.
https://doi.org/10.1007/978-3-030-61143-9_8

have a consecutive syndromes. In 2001, He et al. [9] proposed a new method based on the expression of unknown syndromes according to known ones. They applied it to obtain consecutive syndromes for different QR code lengths. Based on this method, the researchers thought that is possible to apply the very efficient Berlekamp-Massey BM algorithm to the decoding of the QR codes [10–12], because il is an efficient method for determining $L(z)$.

The non-algebraic techniques since the Lookup Table Decoding (LTD) have been applied due to the limitation of the algebraic techniques because they are very complex and use a vast number of operations over a finite field. They accomplishing the same goal and kept the use of the algebraic structure of the QR codes which makes it easy to determine the error patterns in a more direct manner. Furthermore, a hash Table method HT is proposed in [13,14] for decoding systematic QR codes and is mostly used to speed up the LTD by utilizing the hash search scheme. Despite the required memory size and the CPU time for the decoding process, the HT remains an excellent method. Although, all of these previous methods failed to find a single decoding method to decode all QR codes, because each method has different conditions sets for each code.

In this paper, we take into account the challenge of dealing with the difficult decoding of QR codes which make them an inevitable subject. Indeed, we managed to decode, at first, with this method the QR(17, 8, 5) code [1]. In fact, this decoding algorithm is applied to decode the QR(23,12,7) code, while some modifications are adopted in order to apply this method on a large number of QR code. It is important to emphasize that in the proposed decoder, unlike the other, the errors positions can be determined without calculating unknown syndromes and does not need the error locator polynomial $L(z)$, but with simple parameters it seems relatively simple decoding. To validate this method, we have tested all the possible error combinations that do not exceed the correction capacity of the code. The results are very satisfactory since the decoder correct them all. Consequently, we can take into account this method as an effective solution to decode the aforementioned code.

2   Overview of Binary Quadratic Residue Codes

The binary quadratic residue codes are a subordinate class of the cyclic codes, of length $n = 8u \pm 1$, where u is some integer, with a code rate $R \geq 1/2$ and a high minimum distance. To construct a binary QR code $\mathcal{C}(n, k, d)$ of length n, where $k = \frac{n+1}{2}$ and d represent respectively the set of information and the minimal distance, we must find, at first, the smallest positive integer m such that $n | 2^m - 1$. The $\mathbb{F}_{2^m}$ is a finite field of order 2^m. Then, we will define the set Q_n as the collection of non-zero squares elements modulo n as

$$Q_n = \{i | i \equiv j^2 (mod \quad n) \quad \text{for} \quad 1 \leq j \leq n - 1\}. \tag{1}$$

Let α be the root of the primitive polynomial $P(x)$ of degree m and is the generator of the multiplicative group of all the non-zero elements of $\mathbb{F}_{2^m}$, which each element of $\mathbb{F}_{2^m}$ can be expressed by a power of α. Finally, we present the element $\beta = \alpha^{(2^m-1)/n}$ as the primitive n^{th} root of unity in $\mathbb{F}_{2^m}$, which gives rise to the QR code generator polynomial $g(x)$ as

$$g(x) = \prod_{i \in Q_n} (x - \beta^i) \tag{2}$$

For the Golay code the generator polynomial is $g(x) = x^{11} + x^9 + x^7 + x^6 + x^5 + x + 1$. And the codeword $\mathcal{C}$ is defined by the multiplication of the information message and its generator polynomial $g(x)$. Then, Table 1 reunites all the roots and non-roots power of $g(x)$ for QR(23,12,7) code with their corresponding integer values $\Phi(\beta^i)$ and Φ is the so called antilog function.

Table 1. The integer values $\Phi(\beta^i)$ that corresponds to each β^i

i	$\Phi(\beta^i)$	i	$\Phi(\beta^i)$	i	$\Phi(\beta^i)$	i	$\Phi(\beta^i)$	i	$\Phi(\beta^i)$
0	1	5	481	10	1155	15	1525	20	552
1	322	6	637	11	167	16	78	21	1876
2	174	7	1942	12	2011	17	1728	22	1085
3	1164	8	1887	13	418	18	1747		
4	1148	9	1518	14	281	19	319		

3 The Proposed Algorithm for Decoding the Golay Code

Let's assume that a c is a Golay codeword where $n = 23$, $k = 12$, and is transmitted via a noisy channel, which may adversely affect the transmitted codeword by modifying the value of one or more bits. Thus, the received word $r = c + e$ is defined as the sum of the codeword c and the error vector e which represents the channel distortion. Let us define by $r(x) = \sum_{i=0}^{n-1} r_i x^i$ and $e(x) = \sum_{i=0}^{n-1} e_i x^i$, respectively, the polynomial form of the received word and the error where $r(x)$, $e(x) \in \mathbb{F}_2[x]$ and $r_i, e_i \in \{0,1\}$. In fact, if we assume that there are v errors in the received word $r(x)$. The error vector can contain v non-zero terms as

$$e(x) = x^{k_1} + x^{k_2} + \dots + x^{k_v} \tag{3}$$

Where $k_1, k_2, \dots k_v \in \mathbb{Z}_n = [0, n]$ are the error positions to be corrected and $0 \leq k_1 < k_2 < \dots < k_v < n$. Moreover, we define the set of known syndromes which computed directly by evaluating $r(x)$ at the roots of $g(x)$ for all $i \in Q_n$ and we denote the definition of the syndrome S_i as:

$$S_i = r(\beta^i) = e(\beta^i) = \beta^{ik_1} + \beta^{ik_2} + \dots + \beta^{ik_v}. \tag{4}$$

And for the remaining value of i which is not included in the set Q_n, the S_i's are called unknown syndromes.

The decoding process consists of determining the best estimate of $e(x)$ produced by the channel. The difficulty resides in the fact that each QR code has own conditions sets for determining the errors positions. Among this conditions the unknown syndromes set which differs from one code to another. In fact, this means that the QR codes can be decoded efficiently with the same manner once a set of certain unknown syndromes are determined or propose a decoding technique who does not need to calculate the unknown syndromes.

Consequently, we propose a modified non-algebraic decoding algorithm as a second version from the work presented in [1]. In this method, we kept using the algebraic structure of the QR codes and the errors positions can be determined without computing the unknown syndromes nor the error-locator polynomial, hence avoiding the necessity of executing the process of computing a linear equation with high degree. Henceforth, we have created a relationship between the syndromes and the error positions for which the weight goes up to $w(e) = t$. In other words, all error pattern sets are mapped in bijective way into sets of syndromes where each syndrome S_i can be expressed by a sum of power $\beta^{k_1} + \beta^{k_2} + ... + \beta^{k_v}$ with $k_1, k_2, ..., k_v$ correspond to the error positions knowing that $i \in Q_n$. Consequently, we can summarize the possible scenarios in four cases such as (Table 2):

Table 2. The possible scenarios to decode Golay code

$w(e)$	Scenarios	Ratings and relationships
0	Case (0)	$S_i^{(0)} = 0$
1	Case (1)	$S_i^{(1)} = \beta^{ik_1}$
2	Case (2)	$S_i^{(2)} = \beta^{ik_1} + \beta^{ik_2}$
3	Case (3)	$S_i^{(3)} = \beta^{ik_1} + \beta^{ik_2} + \beta^{ik_3}$

Let r be the received words in hard version. First, we assume that a single position is false, which means that $w(e) = 1$. So, we are talking about cases (1). After, evaluating $r(x)$ at the roots of $g(x)$ by using the Eq. (4).

$$S_1^{(1)} = r(\beta) = e(\beta) = \beta^{k'} \tag{5}$$

By calculating the integer value $X^{(1)} = \Phi(S_i^{(1)})$, we search in Table 1 for the exact value $X^{(1)}$ which corresponds to the value obtained in Eq. (5). Thus, k' shows the position of the error. Then, for error weights $w(e) = 2$ or $w(e) = 3$ the things will be more complex and we get the following expressions for $i \in Q_n$:

$$\begin{cases} S_i^{(2)} = \beta^{k'_1} + \beta^{k'_2} \\ S_i^{(3)} = \beta^{k'_1} + \beta^{k'_2} + \beta^{k'_3} \end{cases}$$

Algorithm 1: The proposed algorithm for decoding the Golay code

---

1 Input : Received Codeword r, Table 1
2 Output : Corrected Codeword D
3 **for** $v = 1$ *to* t **do**
4 | $p \leftarrow 0$
5 | **while** $p < n$ **do**
6 | | $i \leftarrow 0$
7 | | **while** $i \in Q_{23}$ **do**
8 | | | compute the syndrome value $S_i^{(v)} \leftarrow r^{(p)}(\beta^i)$ and compute $\Phi(S_i^{(v)})$
9 | | | **if** $\Phi(S_i^{(v)})$ *Exist in Table 1* **then**
10 | | | | correct $r^{(p)}$ and $D^{(v)} \leftarrow r^{(-p)}$ and Break
11 | | | **else**
12 | | | | Look for $X^{(v)}$ and update $r^{(p)}$
13 | | | **end**
14 | | **end**
15 | | $i \leftarrow i + 1$
16 | **end**
17 | $p \leftarrow p + 1$
18 **end**

In fact, we calculate the integer value $\Phi(S_i^{(2)})$, $\Phi(S_i^{(3)})$. In this case, we have seen that the problem is to take from Table 1 the values that contribute to this summation, by mentioning that it is a binary summation.

$$\begin{cases} X^{(2)} = \Phi(S_i^{(2)}) \neq \Phi(\beta^{k'_1}) + \Phi(\beta^{k'_2}) \\ X^{(3)} = \Phi(S_i^{(3)}) \neq \Phi(\beta^{k'_1}) + \Phi(\beta^{k'_2}) + \Phi(\beta^{k'_3}) \end{cases}$$

Then, the most obvious thing is to to check all possible patterns of two β value (respectively three β value), but it does cost a certain amount of complexity which increases relatively as the weight of the error increases and we will fall into the exhaustive decoding. So, we go through Table 1 and we look for the nearest value at X. After that, we modify in r the position k' 'which corresponds to the nearest value at X, then we start the process for $w(e) - 1$. If we cannot decode properly, we move to the second root in Q_{23} and calculate a new syndrome. Moreover, we notice that the algorithm is mainly based on two important nested loops (e.g. line 5 and line 7 in Algorithm 1). The first loop processes $p < n$ circular permutations for received word $r^{(p)} = \pi^{(p)}(r^{(0)})$ where $p = p + 1$, and the second one $r^{(j)} = \pi^{(j)}(r^{(i)})$ processes j circular permutations where $j = (2 \times j) mod\ n$ (i.e $j \in Q_{23}$).

Our aim is to reduce, in the proposed algorithm, the complexity of the decoding process by decreasing a number of permutation used. So, we want to emphasize that this algorithm is written in C language, tested and verified for all possible error patterns. Therefore, it is necessary to check $\sum_{i=1}^{3} \binom{23}{i} = \binom{23}{1} + \binom{23}{2} + \binom{23}{3} = 2047$ error patterns to validate this algorithm of decoding and we obtained good results since the proposed decoder manages to correct

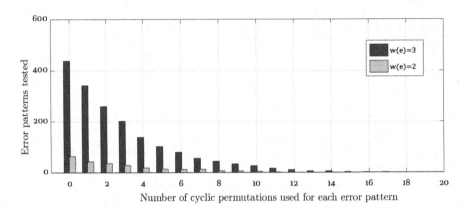

Fig. 1. The eCN that the decoder could have corrected using a $\pi^{(i)}$

all. As result, the Table 3 shows the complexity of the decoder in terms of error patterns number (ePN) that are tested in the worst case. It summarizes the comparison between the pattern test number in the first version of the decoder ePNv1, the modified version ePNv2, the exhaustive search ePNex and the average pattern tested number in our proposed method AePNv2.

Table 3. The decoder complexity in terms of combination tests number

$w(e)$	ePNv1	ePNv2	ePNex	AePNv1	AePNv2
1	1	1	23	1	11,5
2	88	39	253	10.33	126,5
3	2904	1404	1771	243,51	885,5

We note that the method proposed has the smallest value of ePN, it means that the decoder uses a very small number to correct errors of weight $w(e) \leq 3$ and the average value confirms that. In addition, the error patterns tested or decoded did not require the same number of cyclic permutations (cyclic shift). In other words, for a given permutation we have a subset $E_p^{(j)}$ of error patterns where p and j represent respectively the pth permutation and corresponding error weight. We denote by $\Omega^{(j)}$ a set of all error patterns for a given weight j. i.e.

$$\Omega^{(j)} = \bigcup_{p=0}^{n} E_p^{(j)} \tag{6}$$

And, $|E_p^{(j)}|$ is the number of the elements (error patterns) contained in $E_p^{(j)}$ for a giving cyclic permutation p and error weight j. Furthermore, the Fig. 1 presents the number of error patterns that the decoder can corrected using a given cyclic permutation number (cyclic shift) and it is seen that the decoder uses a maximum of 5 cyclic permutations to correct the majority of existing error patterns where the percentage, of the corrected error patterns, amounting to 84%.

Furthermore, in terms of computational complexity, the proposed semi-algebraic decoding algorithm SAD requires simple binary summation and comparisons, unlike the algebraic algorithms that execute an $\mathbb{F}_{2^m}$ arithmetic processors. This suggests that, for QR codes with high lengths, our decoder might be considered an alternative to the BM decoding algorithm and Newton identities.

4 Conclusion

This paper presents a proposed algorithm for decoding the Golay code. This method uses simple parameters to locate the error position without calculating the unknown syndromes or the error locator polynomial and is suitable for both software and hardware implementations. In fact, this decoder can be an alternative way to decode QR codes. For this purpose and by testing all possible error combinations, this can guarantee the robustness and efficiency of our decoder.

References

1. Boualame, H., Chana, I., Belkasmi; M.: New efficient decoding algorithm of the (17, 9, 5) quadratic residue code. In: International Conference on Advanced Communication Technologies and Networking (CommNet), Marrakech, pp. 1–6 (2018)
2. Prange, E.: Some cyclic error-correcting codes with simple decoding algorithms. Air Force Cambridge Res. Center, Bedford, MA, USA, TN-58-156 (1958)
3. MacWilliams, F.J., Sloane, N.J.A.: The Theory of Error-Correcting Codes. North-Holland, Amsterdam (1977)
4. Berlekamp, E. R.: Algebraic Coding Theory, revised edn. McGraw-Hill, Aegean Park Press (1984)
5. Elia, M.: Algebraic decoding of the (23, 12, 7) Golay codes. IEEE Trans. Inf. Theory **33**(1), 150–151 (1987)
6. Chen, X., Reed, I.S., Helleseth, T., Truong, T.K.: Use of Grobner bases to decode binary cyclic codes up to the true minimum distance. IEEE Trans. Commun. **40**(5), 1654–1661 (1994)
7. Reed, I., Yin, X., Truong, T.K.: Algebraic decoding of the (32,16,8) quadratic residue code. IEEE Trans. Inf. Theory **36**(4), 876–880 (1990)
8. Reed, I., Truong, T.K., Chen, X., Yin, X.: The algebraic decoding of the (41 21 9) quadratic residue code. IEEE Trans. Inf. Theory **38**(3), 974–986 (1992)
9. He, R., Reed, I.S., Truong, T.K., Chen, X.: Decoding the (47, 24, 11) quadratic residue code. IEEE Trans. Inf. Theory **47**, 1181–1186 (2001)
10. Chang, Y., Truong, T.K., Reed, I.S., Cheng, H.Y., Lee, C.D.: Algebraic decoding of (71, 36, 11), (79, 40, 15), and (97, 49, 15) quadratic residue codes. IEEE Trans. Commun. **51**, 1463–1473 (2003)
11. Wang, L., Li, Y., Truong, T.K., Lin, T.: On decoding of the (89, 45, 17) quadratic residue code. IEEE Trans. Commun. **61**(3), 832–841 (2013)

12. Li, Y., Liu, H., Chen, Q., Truong, T.K.: On decoding of the (73, 37, 13) quadratic residue code. IEEE Trans. Commun. **62**(8), 2615–2625 (2014)
13. Chen, Y.H., Truong, T.K., Huang, C.H., Chien, C.H.: A lookup table decoding of systematic (47, 24, 11) quadratic residue code. Inf. Sci. **179**, 2470–2477 (2009)
14. Chien, C.H., Huang, C.H., Chang, J.: Decoding of binary quadratic residue codes with hash table. IET Commun. **10**(1), 122–130 (2016)

Iterative Decoding of GSCB Codes Based on RS Codes Using Adapted Scaling Factors

Es-said Azougaghe[1]($\boxtimes$)(iD), Abderrazak Farchane[1], Said Safi[1](iD),
and Mostafa Belkasmi[2](iD)

[1] Polydisciplinary Faculty, Sultan Moulay Slimane University, Beni Mellal, Morocco
essaidazougaghe@gmail.com, a.farchane@gmail.com, safi.said@gmail.com
[2] ENSIAS, Mohammed V University in Rabat, Rabat, Morocco
mostafa.belkasmi@um5.ac.ma

Abstract. In this work, we have extended two algorithms to decode
generalized serially concatenated block codes based on RS codes (GSCB-
RS). The first is the modified Chase-Pyndiah algorithm (MCPA) proposed by Farchane and Belkasmi [1]. The second is the Chase-Pyndiah
algorithm (CPA) that is developed initially for decoding turbo product
codes [2]. We also investigated the effect of different parameters, namely
component codes, the size and structure of the interleaver and the number of iterations, using computer simulations. The simulations result
shows that the performance of the GSCB-RS codes using the MCPA
decoder out performs the CPA decoder that uses predetermined weighting factor (α) and reliability factor (β) parameters.

Keywords: RS codes · Chase decoding · Generalized serial
concatenated block · Modified Chase algorithm · Turbo decoding

1 Introduction

In 1993, Berrou et al. [3] constructed turbo codes. It consists of concatenation
of two recursive convolutional codes separated by a non-uniform interleaver.
These codes showed exceptional performances. One year later, Pyndiah et al. [4]
proposed a new iterative decoding algorithm based on a SISO decoder version of
the Chase decoding. It gives a new directions to researchers in the field. In their
work, they used predetermined parameters weighting factor (α) and reliability
factor (β). The achieved results are similar to those of convolution turbo codes.
But Chase-pyndiah did not decode the concatenated codes (serially, parallel).
The authors [5,6] extended the work of Chase-Pyndiah to decode generalized
concatenated blocks codes using always predetermined parameters α and β.
Moreover, the same authors modified the CPA decoder using adapted parameters
α and β to the context of the decoder. In their work [1] they evaluated the
performance of the MCPA decoder for the product and concatenated codes based
on BCH codes. Our contribution in this paper are extension of the MCPA and

© Springer Nature Switzerland AG 2020
M. Belkasmi et al. (Eds.): ACOSIS 2019, CCIS 1264, pp. 103–114, 2020.
https://doi.org/10.1007/978-3-030-61143-9_9

CPA decoders to decode generalized concatenated block codes based on RS codes, validation of the effectiveness of the MCPA and CPA decoders, and also investigation of same parameters, namely the component codes, the number of iterations, interleaver size and pattern.

This paper is organized as follow: Sect. 2 gives an overview of the encoder structure of the generalized serially concatenated block codes with its classic and new constructions. Section 3 sheds light on the component decoder. In Sect. 4, We present the iterative decoding of the GSCB-RS codes. The simulation results are presented and discussed in Sect. 5. The last section concludes this paper and gives suggestions for further research.

2   Generalized Serially Concatenated Block Codes (GSCB-RS)

Concatenation code is a way to construct large code lengths while keeping their decoding complexity relatively simple. In 1996, Benedetto et al. [7] evaluated the performances of concatenated interleaved codes in a theoretical way, especially SCB codes. However, SCB codes have a weak coding rate, which is not desirable for digital communication systems. Hence in order to overcome this problem, we suggest a new construction allowing to obtain concatenated codes with a very high coding rate. This construction is based on shortcuts codes, it gives us the choice between different coding rates using the same elementary code.

2.1   Classic Construction

The structure of the serially concatenated block codes (SCB) encoder is based on two elementary systematic bloc encoders. The outer $C_1(n_1; k_1)$ code with rate $R_1 = \frac{k_1}{n_1}$ and the inner $C_2(n_2; k_2)$ code with rate $R_2 = \frac{k_2}{n_2}$, with an interleaver π with the length $N = M.n_1$ placed before the second encoder. A block $M.k_1$ bits enters the decoder and is divided to M sub-blocks of k_1 bits each with M representing the number of multi-blocks. Each k_1 is encoded to produce n_1 bit. The encoded blocks are interleaved by an enterleaver before entering the second encoder. $M.n_1$ bits are divided to sub-blocks of n_1 bits which are encoded to produce n_2 bits (Fig. 1).

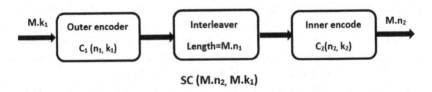

SC (M.n$_2$, M.k$_1$)

Fig. 1. Encoder process of SCB codes

The first encoder produce $p_1 = M.(n_1 - k_1)$ parity bit, and the second produces $p_2 = M.(n_2 - k_2)$ parity bit. So, the sum of parity bits which are generated by the SCB encoder is $p = p_1 + p_2 = M.(n_2 - k_1)$. The code word SCB length is given by $L = M.k_1 + p = M.n_2$. As a result, the encoding rate may be calculated by $R = \frac{M.k_1}{L} = \frac{k_1}{n_2}$. That is to say, the rate of encoding of SCB doesn't depend on the length of the interleaver N By definition, serially concatenated codes have a low rate of encoding, whereas digital systems of communication need a high rate of encoding. To achieve this objective, we used a new construction based on shortcut and symmetric codes.

2.2 New Construction

In this section, we drop the more general case and develop particular cases of serially concatenated block codes using shortening codes. Firstly, we consider serially concatenated block code obtained from two binary linear block code $C_1(n_1, k_1)$ and $C_2(n_2, k_2)$. The outer $\overline{C}_1(n_1 - s_1, k_1 - s_1)$ code is obtained by shortening the code C_1 by s_1 bits. And the inner $\overline{C}_2(n_2 - s_2, k_2 - s_2)$ code is obtained by shortening the code C_2 by s_2 bits, provided that $k_2 - s_2 = n_1 - s_1$, so $s_1 - s_2 = n_1 - k_2$, linked by an interleaver of length $n_1 - s_1$ or $M_1.(n_1 - s_1)$. The overall serially concatenated block (SCB) code is then an $SCB(n_2 - s_2, k_1 - s_1)$ code. The rate of this code is $R = \dfrac{n_2 - s_2}{k_1 - s_1}$.

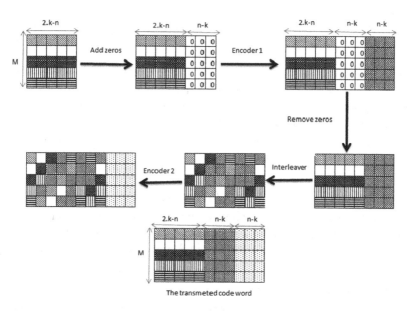

Fig. 2. Encoding process for the GCSB codes

The Fig. 2 illustrates this construction: for each block of $k_1 - s_1$ information bits we add s_1 zeros and encode by $C_1(n_1, k_1)$, then we erase the s_1 zeros from each block. The M blocks are scrambled by an interleaver of length $M.(n_2 - s_2)$.

Again, for each block of $k_2 - s_2$ information bits we add s_2 zeros and encoded by $C_2(n_2; k_2)$, then we erase the s_2 zeros from each block. The code word of the constructed code is formed by the concatenation of the $M.(k_1 - s_1)$ information bits, the $M.(n_1 - k_1)$ parity bits generated by the outer code and the $M.(n_1 - k_1)$ parity bits generated by the inner code.

A special and interesting case for this construction is obtained when $s_1 = s_2 = s$ then $n_1 = k_1$. So, the outer code becomes an $(n_1 - s, k_1 - s)$ code and the inner code becomes an $(n_2 - s, k_2 - s)$. Hence, the rate is given by $R = \frac{k_1 - s}{n_2 - s}$. In particular, if s is equals to $n_1 - k_1$, then the global parameter of the code becomes $SCB(n_2 - n_1 + k_1, 2k_1 - n_1)$. This construction can be used with outer codes of rate greater than one half, and the rate of the code becomes $R = \frac{2k_1 - n_1}{n_2 - n_1 + k_1}$. Now, consider a second interesting case with $s_1 = n_1 - k_2$ and $s_2 = 0$, the later relation implies that $s_1 = n_1 - k_2$, it means that $n_2 - k_2 = n_1 - k_2$. So, the two codes have the same length $n_1 = n_2$. Moreover, we have $s_1 = n_1 - k_1$, this implies that $k_1 = k_2$. The overall serially concatenated block (SCB) code is then an $(n_1, 2k_1 - n_1)$ code. So, $2k_1 - n_1$ should be greeter than zero. The rate of the code becomes $\frac{2k_1 - n_1}{n_1}$, and the interleaver size is k_1 or $M.k_1$.

3   Component Decoder

We consider a transmission over an AWGN channel using BPSK modulation, with code rate k_i/n_i $(i = 1 \text{ or } 2)$. The decoder input equals to $R = C + B$ when the channel is disturbed with AWGN noise knowing that $R = (r_1...r_j...r_{n_i})$ is the observed sequence, $C = (c_1...c_j...c_{n_i})$ $c_j = \pm 1$ is the transmitted codeword while $B = (b_1...b_j...b_{n_i})$ is the vector white noise where the components b_j are zero average and with variance σ^2.

Decoding is realized using the Chase Pyndiah algorithm which allows to determine codewords which are likely. Among these words, it selects those which are at the minimal euclidean distance. The measure of reliability associated to each symbol of the decision d_{if} belonging to the decoded word can be determined starting from the logarithm of the likelihood report LLR bellow:

$$LLR_{if} = \ln \left(\frac{Pr(e_{jf} = +1/R)}{Pr(e_{jf} = -1/R)} \right) \qquad (1)$$

Where e_{if} represents the binary constituent of the transmitted code word E in position (j, f) knowing that $1 \leq j \leq n$ et $1 \leq f \leq m$. The expression of the LLR_{if} can be defined by approximation, in AWGN model where Eq. 1 becomes:

$$LLR_{if} = \frac{1}{2\sigma^2} \left[|R - C^{m(-1)}|^2 - |R - C^{m(+1)}|^2 \right] \qquad (2)$$

Where $C^{m(+1)}$ and $C^{m(-1)}$ are two codewords at minimum euclidean distance from R with $c_{jf}^{m(+1)} \neq c_{jf}^{m(+1)}$, $C^{m(+1)}$ and $C^{m(-1)}$ are selected from the codeword subset given by Chase algorithm. By developing the relation 2 we obtain:

$$LLR_{if} = \frac{2}{\sigma^2} \left(r_{jf} + \sum_{x=1, x \neq j}^{n} \sum_{z=1, z \neq f}^{n} r_{xz} c_{xz}^{m(+1)} \rho_{xz} \right) \qquad (3)$$

Where

$$((x, z) \neq (j, f)) \rho_{xz} = \begin{cases} 0, \text{ if } & c_{xz}^{m(+1)} = c_{xz}^{m(-1)} \\ 1, \text{ if } & c_{xz}^{m(+1)} \neq c_{xz}^{m(-1)} \end{cases}$$

by normalizing the approximated LLR of d_{if} with respect to $\frac{2}{\sigma^2}$ we obtain:

$$r'_{jf} = (\frac{\sigma^2}{2}).LLR_{if} = r_{jf} + w_{jf} \tag{4}$$

The LLR of a symbol is equal to the sample sum represented at the decoder input and an amount w_{jf} which is independent of r_{jf} and analogous to the extrinsic information of the convolutive turbo-codes. To determine the LLR simplified expression of the output symbol, it is necessary to determine the two code words at the minimum distance from R and having two symbols of opposite signs $(j, f)(i = \pm 1)$, to do this, we use Chase algorithm which allows us to determine a subset of code words from which we can find the two words we are looking for. $C^{m(+i)}$ must have $-i$ as binary element as position (j, f).

If the $C^{m(+i)}$ codeword is found, the soft decision r_{jf} can be computed using the relation given below:

$$r'_{jf} = \left(\frac{(M^{m(-i)} - M^{m(i)})}{4} \right) c_{jf}^{m(i)} \tag{5}$$

Where $M^{m(-i)}$ and $M^{m(i)}$ represent respectively the $c_{jf}^{m(i)}$ euclidean distance from R and $c_{jf}^{m(i)}$ euclidean distance from R.

Sometimes, we don't find, in the subset, two words with a minimum distance R having symbols with opposite signs. In this case, the LLR of the chosen symbols is given by this relation:

$$r'_{jf} = \beta c_{jf}^{m(i)} \tag{6}$$

where β is a constant which is a function of the iteration.

4   Decoding Process of GSCB-RS Codes

4.1   GSCB-RS Decoder

The structure of the GSCB decoder is shown in Fig. 3.

To decode a wordcode of GSCB code, we will need not only extrinsic information on the information bits but also extrinsic information on the redundancy bits, unlike GPCB codes [8], we will need extrinsic information on the information bits. The GSCB decoder is constructed from two elementary decoders, the Inner and Outer decoder.

The first decoder is similar to that of the GPCB decoder [5]. The second is a little different from the first. Initially, the elementary decoder 1 has no knowledge

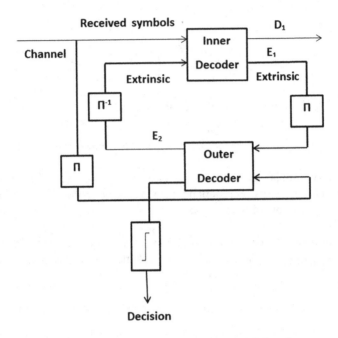

Fig. 3. Decoding process for the GSCB-RS codes

on the information bits except the information observed on the channel. Then, it computes extrinsic information on the information bits using the decoder from Chase-Pyndiah. This extrinsic information is deinterleaved and feeds the elementary decoder 2 together with the sequence received by the channel. This latter generates extrinsic information on information bits and redundancy bits. This procedure determines an iteration. The extrinsic information obtained by the second decoder is interleaved, then feeds the first decoder if the iterations are to continue. This process starts again until a maximum number of iterations is reached.

The Chase-Pyndiah decoder uses two factors weighting factor (α) and reliability factor (β), the choice of these two factors is determined experimentally. This process is described below. The values obtained by this process are shown in Table 1.

4.2   Interleaver

Interleaving or permutation functions, used between elementary coders in a concatenated scheme, have two roles. On the one hand, they ensure, at the output of each elementary decoder, a temporal dispersion of the errors which can be produced in bursts. These error packets then become isolated errors for the following decoder, with much less correlation effects. This error dispersion technique is used in a broader context than that of channel coding. We use it with profit for example to reduce the effects of the long attenuations in transmissions affected

Table 1. Simulation parameters

Parameter name	Value
Modulation	BPSK
Environment	The C programming language
Channel	AWGN
Interleaver structure	Random (default value)
	Cyclic
	Primitive
	Diagonal
	Helical
α	0.01, 0.07, 0.09, 0.13, 0.15, 0.19, 0.25, 0.29, 0.33, 0.37, 0.41, 0.45, 0.49, 0.53, 0.57, 0.61, 0.63, 0.69, 0.73, 0.75
β	0.51, 0.55, 0.60, 0.65, 0.70, 0.74, 0.78, 0.80, 0.83, 0.85, 0.87, 0.89, 0.90, 0.92, 0.95, 0.96, 0.97, 0.98, 0.99, 1.00
Elementary decoder	Chase-Pyndiah
Iterations	From 1 to 15 (default)
Interleaver size	$1 \times k$, $10 \times k$, $100 \times k$, $300 \times k$

by fading, and generally in situations where disturbances can alter consecutive symbols. On the other hand, in close connection with the characteristics of the component codes, the permutation is designed so that the minimum distance of the concatenated code is as great as possible.

4.3   Choice of Empirical Parameters α and β

α and β parameters are empirically determined [9,10]. These parameters play an important role in yielding better performance when we choose a medium length code, and we take a relatively high parameter value M (for example $M = 100$). We consider these parameters to be optimal because we have made a compromise between the length of the code and the computational complexity. We start our trial process by setting the number of iterations to 1, and we change the value of α, with $0 < \alpha < 1$. We retain the value of α giving the best performance. Then we change, in the same way, the value of β, with $0 < \beta < 1$. Once the values of α and β are chosen for the first iteration, we increment the number of iterations, and we seek the values of α and β for the second iteration. After that, we go back, without decreasing the number of iterations, to adjust the values of α and β for possible performance improvements. Finally, we increment the number of iterations and repeat the same process until the last iteration.

4.4   Adapted Parameter α and β

Parameter $\alpha(p)$: In several studies [4,5,9,11,12], the parameter is chosen empirically, and its value gives a $BER = 10^{-5}$ with a minimum of iterations for a given SNR. This empirical method is tedious and time consuming. To solve this problem, we have adapted this parameter for product codes and generalized concatenated block codes. The following equation gives the expression of $\alpha(p)$:

$$\alpha(p) = \frac{1}{\sigma^2_{W(p-1)}} \tag{7}$$

where $\sigma^2_{W(p-1)}$ denotes the variance of the normalized extrinsic information of the code word delivered by the previous decoder. The performances attained using the adapted $\alpha(p)$ parameter are comparable to the ones obtained by the predetermined parameter.

We have used the adapted $\alpha(p)$ parameter 7 to evaluate the performance of the GSCB-RS(63,39) codes. The results obtained by this parameter are comparable to those obtained by the predetermined alfa parameter.

We used the adapted $\alpha(p)$ parameter (7) to evaluate the performance of the GSCB-RS(63,39) codes. The Fig. 4 shows the performance obtained using the advantage of adapted parameters is that they do not need to be re-optimized if we change the application. However, the predetermined parameters need to be re-optimized when we change the code or modulation.

Parameter β: The β parameter is used in the following situation: if the competitor is absent, all the code words have an element c_j equal to d_j. This means that all code words have the same decision about the element d_j. Therefore, the LLR at the decoder output should confirm this decision. In this case, the reliability produced by the decoder must follow the fact that all the words agree on the same decision d_j. The normalized LLR noted by γ_{d_j} can be translated by the following relation:

$$\gamma_{d_j} = \beta.d_j \tag{8}$$

where

$$\beta = (\sigma_\lambda + |\lambda_j|)$$

where σ_λ is the standard deviation of the decoder input sequence R.

5   Simulation and Results

This section presents the performance of the GSCB-RS code on an additive white Gaussian noise (AWGN) transmission channel. During this simulation, we used a BPSK modulation. We focus on the bit error rate (BER) in relation to the signal to noise ratio (SNR). The BER is influenced by numerous parameters, namely the elementary code used, the number of iterations, the structure and the size of the interleaver.

The Fig. 4 depicts the performances of the code GSCB-RS(63, 39). Here, we fix the multi-block parameter at the value 300. From this figure, we notice

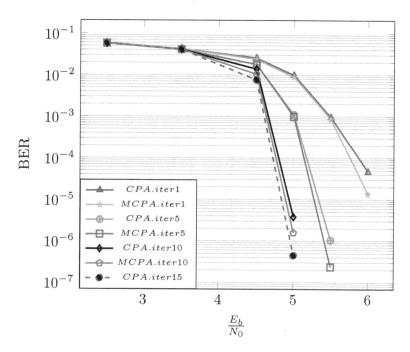

Fig. 4. Iterations effect on the performance of GSCB-RS(63, 39) Code, with M = 300

that the BER improves with iterations. Nevertheless, after the 10th iteration the coding gain becomes negligible for the CPA decoder on the other hand the MCPA decoder can go up to the 15th iteration.

The Fig. 5 shows the effect of the multi-block M. The gain reaches 1.4 dB as we pass from $M = 1$ to $M = 10$, decreases to 0.4 dB between $M = 10$ to $M = 100$ and becomes negligible beyond $M = 100$. This demonstrates how effective is the multi-block M.

In order to evaluate the effect of the interleaver structure on the performance of GSCB-RS codes, we represent in the same figure the BER against SNR of the code GSCB-RS(63,39) for different interleavers, namely the random, primitive, cyclic, diagonal and helical interleaver. It should be noted that the multi-block used here is M = 300.

The Fig. 6 depicts the performance results. According to this figure we observe that the Random interleaver outperforms the other ones by about 0.5 dB at BER = 10^{-5}.

In this part, a comparison of the performance of the GSCB-RS (127.85) and GSCB-RS (63.39) codes is conducted. The last two codes have similar coding rate, whose value R equals 0.82 and the number of multi-blocks used M equals 300. The performance of the later codes is presented in Fig. 7. From this results, we observe that when we increase the component code length the performance becomes worse. The GSCB-RS (63,39), GSCB-RS (127, 85) codes are respectively 2.1, 2.6 away from their Shannon limit.

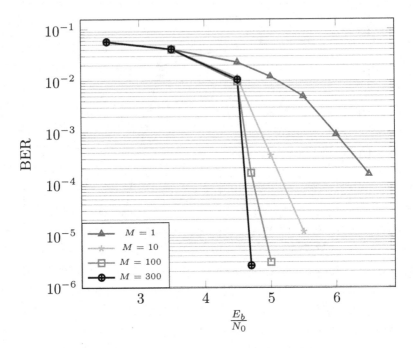

Fig. 5. Multi-block effect on the performance of GSCB-RS(63, 39) code

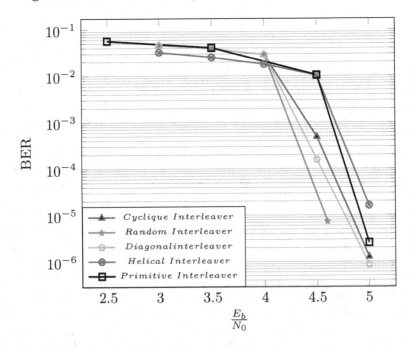

Fig. 6. Interleaver structure effect on the performance of GSCB-RS(63, 39) Code, M = 300

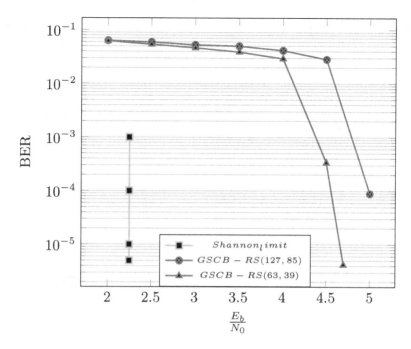

Fig. 7. Comparison between the performance of $GSCB - RS(127, 85)$ and $GSCB - RS(63, 39)$ codes

6   Conclusion

This work extend the MCPA and CPA decoders to decode GSCB-RS codes. The simulation result demonstrates the effectiveness of the decoder MCPA proposed in [1], and shows that the MCPA decoder out performs the CPA decoder. The MCPA decoder can be extended to decode product and hybrid concatenated block codes based on RS and BCH codes.

References

1. Farchane, A., Belkasmi, M.: New efficient decoder for product and concatenated block codes. J. Telecommun. **12**, 17–22 (2012)
2. Pyndiah, R.M.: Near-optimum decoding of product codes: block turbo codes. IEEE Trans. Commun. **46**(8), 1003–1010 (1998)
3. Berrou, C., Glavieux, A., Thitimajshima, P.: Near Shannon limit error-correcting coding and decoding: Turbo-codes. 1. In: 1993 IEEE International Conference on Communications. ICC 1993 Geneva, Technical Program, Conference Record, vol. 2, pp. 1064–1070. IEEE (1993)
4. Pyndiah, R., Glavieux, A., Picart, A., Jacq, S.: Near optimum decoding of product codes. In: GLOBECOM94 (1994)

5. Belkasmi, M., Farchane, A.: Iterative decoding of parallel concatenated block codes. In: IEEE International Conference on Computer and Communication Engineering, 2008. ICCCE 2008, pp. 230–235 (2008)

6. Farchane, A., Belkasmi, M.: Generalized serially concatenated codes: construction and iterative decoding. Int. J. Math. Comput. Sci. **6**(2) (2010)

7. Benedetto, S., Divsalar, D., Montorsi, G., Pollara, F.: Serial concatenation of interleaved codes: performance analysis, design, and iterative decoding. IEEE Trans. Inf. Theory **44**(3), 18 (1998)

8. Azougaghe, E., Farchane, A., Safi, S., Belkasmi, M.: Adapted scaling factors for decoding concatenated codes based on RS codes. Int. J. Adv. Sci. Technol. **126**(1), 1–10 (2019)

9. Farchane, A., Belkasmi, M., Nouh, S.: Generalized parallel concatenated block codes based on BCH and RS codes, construction and Iterative decoding. J. Telecommun. **12** (2012)

10. Ayoub, F., Farchane, A., Askali, M., Belkasmi, M., Himmi, M.M.: Serially concatenated OSMLD codes: design and iterative decoding. Appl. Math. Sci. **10**, 2179–2188 (2016)

11. Aitsab, O., Ramesh, P.: Performance of concatenated Reed-Solomon/convolutional codes with iterative decoding. In: IEEE Global Telecommunications Conference, GLOBECOM 1997, vol. 2, pp. 934–938. IEEE (1997)

12. Askali, M., Ayoub, F., Chana, I., Belkasmi, M.: Iterative soft permutation decoding of product codes. Comput. Inf. Sci. **9**(1), 128–135 (2016)

Turbo Decoder Based on DSC Codes
for Multiple-Antenna Systems

Abdelghani Boudaoud$^{(\boxtimes)}$, Mustapha El Haroussi,
and Elhassane Abdelmounim

FST, Hassan I University, Settat, Morocco
Boudaoud586@gmail.com, m.elharoussi@gmail.com,
hassan.abdelmounim@hotmail.fr

Abstract. In this paper, we present the contribution of the insertion of a Turbo-type channel decoder in a MIMO chain. This MIMO chain is based on Orthogonal Space-Time Block Code (OSTBC). Alamouti proposed two structures, based on the OSTBC code and having two transmitting antennas: the first structure has a single receiving antenna, that is OSTBC 2×1 and the second one has two receiving antennas that is OSTBC 2×2. The turbo-decoder is based on the Difference Set Codes-One Step Majority Logic Decodable (DSC-OSMLD); it is the DSC (21, 11) code. After the introduction of this turbo decoder in the two Alamouti's structures, performance are noted and compared in terms of the Bit Error Rate (BER) versus the Signal-to-Noise Ratio (SNR). The obtained results show that the addition of a one receiving antenna to the 2×1 OSTBC structure provides a decoding gain equal to 1 dB, while the insertion of the proposed turbo decoder brings a gain of 5.5 dB at the first iteration only.

Keywords: Bit Error Rate · DSC code · Errors correcting codes · MIMO · OSTBC code · Turbo decoding

1 Introduction

MIMO transmission techniques consist in using several transmit and receive antennas. What aims to form multi-paths and exploit them to increase performances in terms of capacity, reliability and spectral efficiency.

S.M. Alamouti [1] presented, in 1998, the scheme of a diversity system that increases the reliability of the transmission for two transmit antennas. In 1999, Tarokh [2] extended the Alamouti coding for an arbitrary number of antennas. The Alamouti and Tarokh codes are orthogonal Space-Time Block Codes (OSTBC).

MIMO systems have become essential elements in many wireless communication standards. Among these standards we can cite as examples: Wi-Fi (IEEE 802.11n and IEEE 802.11ac standards) which offers a flow rate exceeding 1 Gbps in mode MIMO 4×4, whereas the previous versions do not reach the 100 Mbps (54 Mbps for IEEE 802.11a/802.11 g standards) [3].

The notion of turbo-codes was discovered by C. Berrou [4] in 1993, it is considered as an essential advance in information transmission systems. Indeed, most transmission

© Springer Nature Switzerland AG 2020
M. Belkasmi et al. (Eds.): ACOSIS 2019, CCIS 1264, pp. 115–120, 2020.
https://doi.org/10.1007/978-3-030-61143-9_10

standards have adopted these codes. They are also used in ADSL-2 [5], in the 4G-LTE and LTE-Advanced mobile networks [6].

In this work, we propose to insert a turbo decoder, based on a DSC code of small length DSC (21, 11) [7], in a MIMO structure based on the OSTBC code. We will study the Bit Error Rate (BER) performances, and analyze them for two structures. The first one is 2×1: two transmitting antennas and one receiver, and the second is 2×2: two transmitting antennas and two receivers.

This paper is organized as follows: in Sect. 2 and 3 we will present, in order, the principle of Turbo decoding and that of MIMO systems using the OSTBC code. Section 4 is reserved for experimentation and discussion of results.

2   The Turbo-Decoding

2.1   Product Codes

The first turbo codes were based on convolutional codes. In 1994 R. Pyndiah [8] proposed the turbo codes in block (TCB), these TCBs use weighted input and output decoding (SISO: Soft In-Soft Out). In this work, we have used an iterative decoding process that follows the model proposed by Pyndiah [8] and builds on "One Step Majority Logic Decodable, Difference Set Codes" (OSMLD-DSC) [9, 10], using the soft-out extension of the Massey threshold decoding [7, 11–14].

Let there be two systematic linear block codes $C_1(n_1, k_1)$ and $C_2(n_2, k_2)$ where n_i and k_i are, in this order: the length of the code, and the number of symbols of the information ($i = 1$ or 2). From the two codes C_1 and C_2, it is possible to construct a product code $C_P = C_1 \otimes C_2$, which has the parameters $(n_1 n_2, k_1 k_2)$. The C_P code is obtained as follows: we code the lines of a matrix of $k_2 \otimes k_1$ information symbols by C_1, and we obtain the matrix constituted from $k_2 \otimes n_1$ symbols. Then we code the columns of this last matrix by C_2. In this work, we have used for C_1 and C_2 the same DSC code with small dimensions. Hence $C_1 = C_2 = $ DSC (21, 11). The performances of the turbo decoder based on this code are studied in the reference [7].

2.2   Turbo Decoding Principle

A turbo decoder consists of SISO decoders (generally two decoders) and interleavers (see Fig. 1). The symbols are received from the channel, line by line, by the first decoder. This decoder gives the information a priori soft, and then the information about the channel and the extrinsic information are interleaved and provided, column by column, to the second decoder. A complete iteration consists of activating of each decoder once. So perform more iterations, the decoder will converge towards the correct solution; however more iterations often require more time [7, 14–16].

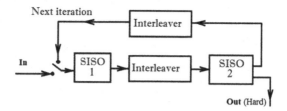

Fig. 1. Turbo decoder structure.

3 Principle of MIMO (OSTBC) System

Let send two symbols S_0 and S_1, and T the duration of transmission of a symbol. The first scheme of Alamouti is based on two transmit antennas, and uses the OSTBC coding. The first transmission antenna transmits the symbol S_0 then $(-S_1^*)$. Simultaneously, the second antenna transmits the symbols S_1 and then (S_0^*). In its basic form, the Alamouti scheme [5] requires a single receiving antenna as shown in Fig. 2. If it is assumed that the channel is invariant between the first and the second transmission time, as shown in expressions (1), then the received signal r, at times t and (t + T), can be expressed by the system (2). With n_0 and n_1 are complex random variables representing the noise of the receiver and H_0 and H_1 are the transmission coefficients of the channel.

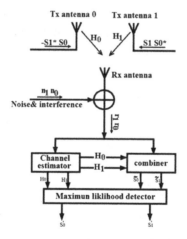

Fig. 2. Basic scheme of Alamouti's OSTBC with a single receiver [1]

$$\begin{cases} H_0 = H_0(t) = H_0(t+T) = a_0 e^{j\theta_0} \\ H_1 = H_1(t) = H_1(t+T) = a_1 e^{j\theta_1} \end{cases} \tag{1}$$

$$\begin{cases} r_0 = r(t) = s_0.H_0 + s_1.H_1 + n_0 \\ r_1 = r(t+T) = -s_1^*.H_0 + s_0^*.H_1 + n_1 \end{cases} \tag{2}$$

The main advantage of this coding is that the signal can be recovered using a simple linear operation without amplifying the noise [1], the values estimated by the Alamouti combination scheme of the signals S_1 and S_0 are:

$$\begin{cases} \tilde{s}_0 = H_0^*.r_0 + H_1 r_1^* \\ \tilde{s}_1 = H_1^*.r_0 - H_0 r_1^* \end{cases} \quad \text{or}$$

$$\begin{vmatrix} \tilde{s}_0 \\ \tilde{s}_1 \end{vmatrix} = \begin{vmatrix} H_0^* & H_1 \\ H_1^* & -H_0 \end{vmatrix} \times \begin{vmatrix} r_0 \\ r_1^* \end{vmatrix} \tag{3}$$

In the same reference [1] we will find more details about the principle of OSTBC 2×2 coding.

4   MIMO System with Turbo Decoder

4.1   Chain of Simulation

The Fig. 3 shows the schematic of the chain that we have designed for measuring the BER performance versus SNR. In this figure we can see: a Bernoulli binary generator, which provides a random sequence of bits (0 and 1), an OSTBC encoder and combiner, a simulator of the MIMO channel, the product encoder and the turbo decoder.

In the first time, we designed this platform without "product encoder" and "turbo decoder", in aim to measure the performance of OSTBC structures 2×1 and 2×2 without turbo-decoding. At the second time, we introduced into the chain the different blocks related to the Turbo decoding.

The product code used is based on DSC $(21, 11) \otimes$ DSC $(21, 11)$ code, for more information see reference [7].

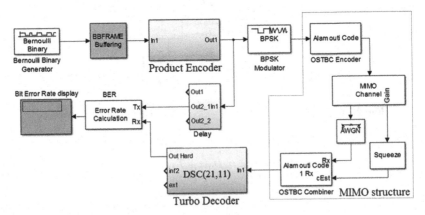

Fig. 3. Diagram of the MIMO simulation chain with the turbo decoder on Matlab/Simulink software

4.2 Performances BER vs SNR

The study examines the BER vs SNR performances of two Alamouti MIMO structures: OSTBC 2 × 1 and OSTBC 2 × 2. For each of the structures we considered three configurations: OSTBC alone, OSTBC with Turbo decoder at a single iteration and OSTBC with Turbo decoder at two iterations (see Fig. 4). The results obtained show, on the one hand, that the increase in the number of antennas used improves the BER performance of the chain. Especially for a given configuration, there is a gain of about 1 dB between a configuration using the 2 × 1 structure and the same configuration using the 2 × 2 structure. On the other hand, for a given structure, turbo decoding brings a gain of 5.5 dB at the first iteration and 7.5 dB at the second iteration.

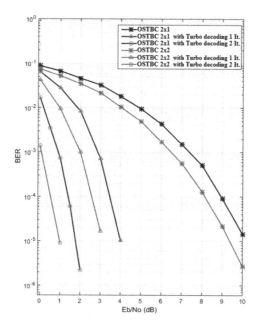

Fig. 4. The BER performances of Alamouti schemes with turbo decoding.

5 Conclusion

The results, obtained in this work, show that the introduction of the channel decoder, and particularly the turbo-decoder, into the MIMO structures is an effective solution for the correction of transmission errors, even in environments that are too noisy. We have shown that the addition of one receiving antenna to the 2 × 1 OSTBC brings a gain of 1 dB, while the insertion of the proposed turbo decoder brings a gain of 5.5 dB at the first iteration only. The turbo decoder used is based on a code of short length, the DSC (21, 11). The FPGA implementation of this turbo decoder had been studied in reference [7], where it had been shown that it's characterized by low latency, low complexity and high throughput.

In the future work, we will introduce the decoder, implemented in FPGA circuit, in MIMO structure, and we will investigate its performances.

References

1. Alamouti, S.M.: A simple transmit diversity technique for wireless communications. IEEE J. Sel. Areas Commun. **16**(18), 1451–1458 (1998)
2. Tarokh, V., Jafarkhani, H., Calderbank, A.R.: Space-time block codes from orthogonal designs. IEEE Trans. Inf. Theory **45**(15), 1456–1467 (1999)
3. IEEE Computer Society: Part 11: Wireless LAN Medium Access Control (MAC) and Physical Layer (PHY) Specifications (2007)
4. Berrou, C., Glavieux, A., Thitimajshima, P.: Near Shannon limit error correcting coding and decoding: Turbo-Codes. In: IEEE International Conference on Communication ICC 1993, vol. 3, no. 12, pp. 1064–1070, May 1993
5. Michel, H., Wehn, N.: Turbo-decoder quantization for UMTS. Commun. Lett. IEEE **5**(12), 55–57 (2001)
6. Khan, F.: LTE for 4G Mobile Broadband. Cambridge University Press, New York (2009)
7. Boudaoud, A., El Haroussi, M., Abdelmounim, E.: VHDL design and FPGA implementation of a high data rate turbo decoder based on majority logic codes. Int. J. Electr. Comput. Eng. (IJECE) **7**(13), 1824–1832 (2017)
8. Pyndiah, R., Glavieux, A., Picart, A., Jacq, S.: Near optimum decoding of product codes. In: Globecom 1994, San Francisco, vol. 1, no. 13, pp. 339–343, November–December 1994
9. Weldon Jr., E.: Difference-set cyclic codes. Bell Syst. Techn. J. **45**, 1045–1055 (1964)
10. Rudolph, L.: Geometric configuration and majority logic decodable codes, Norman (1964)
11. Massey, J.: Threshold Decoding. MIT Press, Cambridge (1963)
12. Graham, R.L., MacWilliams, F.J.: On the number of information symbols in difference-set cyclic codes. Bell Syst. Tech. J. **45**, 1057–1070 (1966)
13. MacWilliams, F., Seery, J.: The weight distributions of some minimal cyclic codes. IEEE Trans. Inform. Theory **1**(2IT-27), 796–806 (1981)
14. Elharoussi, M., Belkasmi, M.: VHDL design and FPGA implementation of a fully parallel architecture for iterative decoder of majority logic codes for high data rate applications. J. Wirel. Netw. Commun. (JWNC) **2**(14), 35–42 (2012)
15. Santos, I., Murillo-Fuentes, J.J.: Self and turbo iterations for MIMO receivers and large-scale systems. IEEE Wirel. Commun. Lett. **8**(4), 1095–1098 (2019)
16. Bakulin, M., Kreyndelin, V., Rog, A., Petrov, D., Melnik, S.: A new algorithm of iterative MIMO detection and decoding using linear detector and enhanced turbo procedure in iterative loop. In: 24th Conference of Open Innovations Association (FRUCT), Moscow, Russia, pp. 40–46 (2019)

A Threshold Decoding Algorithm for Non-binary OSMLD Codes

Zakaria M'rabet[1](✉) ⓘ, Fouad Ayoub[2]ⓘ, and Mostafa Belkasmi[1]ⓘ

[1] ICES Team, ENSIAS, Mohammed V University in Rabat, Rabat, Morocco
zakaria.mrabet@um5s.net.ma
[2] ERIATC Team, LaREAMI Lab, CRMEF-K, Kenitra, Morocco

Abstract. In this paper, a non-binary version of the Soft-Input Hard-Output Threshold Decoder (TD) is proposed. Threshold decoding was initially devised by Massey for the binary case. This work is the continuity of our previous work, where we presented the Majority Logic Decoding (MLGD) algorithm for non-binary codes. Like the (MLGD), the proposed decoder has low complexity, but it is more suited for OSMLD codes with a large orthogonal structure in their dual codes, such as finite geometry codes. The performances of our decoder are shown in terms of bit, symbol, and block error rates. The obtained results are very encouraging. The obtained results of the TD show an amelioration compared to the MLGD, due to the use of the information received from the channel.

Keywords: Galois Field (GF) · Finite euclidean geometry (EG) · Non-binary codes · One-step majority logic decodable (OSMLD) codes · Threshold decoding (TD) · Finite geometry LDPC (FG-LDPC) codes

1 Introduction

Majority-logic decoding is a simple and effective scheme for decoding certain classes of block codes, especially for decoding certain classes of cyclic codes. The first majority-logic decoding algorithm was devised in 1954 by Reed [15] for the class of Reed-Muller (RM) codes. Reed's algorithm was later extended and generalized by coding theorists. The first unified formulation of the majority-logic decoding was due to Massey [7]. Most majority-logic decodable codes found so far are cyclic codes [6]. Important cyclic codes of this category are codes constructed based on finite geometries, namely, Euclidean and projective geometries. These codes are called finite geometry codes and they contain punctured RM codes in cyclic form as a subclass [6].

Finite geometry codes were first investigated by Rudolph [16] in 1964. Rudolph's work was later extended and generalized by many coding researchers, from the late 1960s to the late 1970s [3,5,8,13], to name a few. The finite geometry codes were rediscovered in 2000 by Lin and Fossorier [4], as they showed that a special subclass of finite geometry codes forms a special subclass of good

© Springer Nature Switzerland AG 2020
M. Belkasmi et al. (Eds.): ACOSIS 2019, CCIS 1264, pp. 121–133, 2020.
https://doi.org/10.1007/978-3-030-61143-9_11

low-density parity-check (LDPC) codes. Therefore the finite geometry codes can be considered classical and modern at the same time.

Decoding binary finite geometry codes by MLGD can easily be implemented in hardware with a feedback shift register and a majority-logic circuit. This hardware implementation can achieve very high decoding speed and is quite suitable for high-speed optical networks operating at 10 Gbits or for high-speed satellite communications [6]. Even though decoding non-binary finite geometry codes with MLGD requires more complexity depending on the field alphabet GF(q), it is still much easier than other known non-binary decoder, as we will discuss later.

Non-binary Low-Density Parity Check (LDPC) codes have received considerable attention since Davey and McKay devised their first version of q-ary sum-product algorithm (QSPA) in 1998 [17], which is a message-passing algorithm based in belief propagation. The QSPA is too heavily complex, which makes it not suitable for practical applications. After several ameliorations, the complexity problem con. Therefore new decoders emerged based on Majority Logic Decoding, which is in a way similar to the belief propagation but works with much less complexity. The problem of these decoders is that they don't suit codes with small column weights in their parity check matrices.

In our first work [18], the MLGD for non-binary codes was investigated. This algorithm was the lightest in terms of decoding complexity but was more suitable for one-step majority logic decodable OSMLD codes, especially those of large orthogonal structure. Fortunately, the family of finite geometry codes has relatively large column weight, and due to the geometrical characteristics of their construction, their column weight represents also the number of vectors involved in the orthogonal structure. The greater the orthogonal structure, the more information the decoder will have to decide on a symbol. The Euclidean Geometry (EG) and Projective Geometry (PG) codes are the most famous of the finite geometry LDPC codes, and they work perfectly for the decoders based Majority Logic. The reason for the low complexity of the MLGD is that it consists of a majority vote of a set of orthogonal equations to estimate each data symbol.

To continue in the same direction, our aim in this paper is to produce a Soft-in Hard-out decoder based majority-logic decoding. To do so, we need to incorporate the information received from the channel, while keeping a relatively low complexity compared to known non-binary decoding algorithms. The reason for which we chose to develop a non-binary version of the Threshold Decoding (TD), by simplifying some formulas. The binary version of this latter was first proposed by Massey in 1963 [7], which is less complex than the binary Sum-Product Algorithm (SPA).

Since decoding complexity of non-binary codes is a key challenge, in a previous work we derived a non-binary version of the MLGD, which is a hard-input hard-output decoder. In this paper we upgraded it to a soft-input hard-output decoder, to benefit from the information at the channel output, which tends to improve the performance of the new decoder. This paper is organized as follows. In the second section, we give a preliminary review of non-binary OSMLD error-correcting codes. Then, in Section III, we describe the proposed decoder, which is most suitable for EG and PG codes. In section IV, we exhibit the performance of our decoder in terms of bit error rate (BER), symbol error rate (SER) and block error rate (BLER), then, we analyze the computational complexity of our decoder. The last section will conclude the paper.

2 Non-binary OSMLD Euclidean Geometry Codes

2.1 Preliminaries and Notations

Error-correcting codes are used to reduce the errors in a received sequence of symbols that was transmitted through a noisy channel. In the non-binary codes, the transmitted symbols belong to a Galois Field, and each symbol in a Galois field has a binary representation. Let α be the primitive element of a Galois Field $GF(q)$, where $q = 2^m$ and m a positive integer. Then $GF(q) = \{\alpha^{-\infty} = 0, \alpha^0 = 1, \alpha, \alpha^2, \ldots, \alpha^{q-2}\}$, and each element α_s has a binary representation on m bits, with $s = 0, 1, \ldots, q - 1$. Therefore $\alpha_s \in GF(q)$, $\alpha_s = (a_{s,0}, a_{s,1}, \ldots, a_{s,m-1})$ is its binary representation.

Consider a non-binary cyclic code C with (n, k, d_{min}) representing respectively its code length, dimension and minimal distance. Let $GF(q)$ be the field alphabet of the code C with $q = 2^m$, and let $\underline{c}$ be a codeword, i.e. $\underline{c} \in GF(q)^n$. Let $H = [h_i]$ with $i = 1, \ldots, M$ be the parity check matrix, whose component also belong to $GF(q)$, and where h_i is its ith row. In general for LDPC codes $M > n - k$. The row space of H is an $(n - k, n)$ cyclic code, denoted by $C^\perp$ or C_d, which is the dual code of C, or also called the null space of C. For any vector $\underline{c} \in C$ and $\underline{w} \in C^\perp$, the inner product of $\underline{c}$ and $\underline{w}$ is 0 in $GF(q)$, that is,

$$\underline{c}.\underline{w} = c_1 w_1 + c_2 w_2 + \cdots + c_n w_n \tag{1}$$

In fact, an n-tuple $\underline{c}$ is a code-word in C means for any vector $\underline{w}$ of the dual code $C^\perp$, $\underline{c} \cdot \underline{w} = 0$. The equality (1) is called a *parity check* equation. It is clear that there are 2^{n-k} such parity check equations. We assume a transmission over an Additive White Gaussian Noise (AWGN) channel using a Binary Phase shift Keying (BPSK) modulation such that each symbol c_j is first expanded into its binary form $c_j = (c_{j,0}, c_{j,1}, \ldots, c_{j,m-1})$ then modulated following the rule in (2):

$$x_{j,b} = (-1)^{c_{j,b}} \tag{2}$$

Therefore the modulated codeword $x = \{-1, 1\}^{mn}$. The transmission is performed assuming the model $y = x + e$, where $y \in \mathbb{R}^{mn}$ is the received signal, and $e \in \mathbb{R}^{mn}$ is the additive white Gaussian noise vector with zero mean and variance $N_0/2$, where N_0 is the noise spectral density.

Let $z' = (z'_{1,1}, \ldots, z'_{mn}) \in GF(2)^{mn}$ denote the binary hard decision of the received signal y, defined by:

$$z'_{j,b} = (1 - sgn(y_{j,b}))/2 \tag{3}$$

Then by grouping m-by-m the bits in z', we obtain the non-binary hard decision vector $z = (z_1, \ldots, z_n)$, which will be used in the decoding process.

Usually, to detect the occurrence of an error within a transmitted codeword, we compute its syndrome $\underline{s} = \underline{z}.H^T$, and if $\underline{s} = \underline{0}$ then $\underline{z}$ is a codeword in C.

Let $\underline{z}$ be the hard decision received vector, therefore for any vector $\underline{w}$ in the dual code $C^{\perp}$, we can form the following linear sum:

$$A = (\underline{z}.\underline{w}).w_n^{-1} = (z_1 w_1 + z_2 w_2 + \cdots + z_n w_n).w_n^{-1} \tag{4}$$

which is called the *parity check sum* or simply *check sum*. This parity check A must be equal to zero if the received vector $\underline{z}$ is a code-word in C, however, if $\underline{z}$ is not a code-word, then A may not be equal to zero. A received symbol z_j is said to be *checked* by the check sum A if the coefficient w_j is non-zero.

Since the code is linear, normalizing the vector $\underline{w}$ by the last symbol w_n^{-1} will result in another vector $\underline{w}'$ whose last component $w_n' = 1$, therefore (4) becomes:

$$A = \underline{z}.\underline{w}' = z_1 w_1' + z_2 w_2' + \cdots + z_n \tag{5}$$

Since $\underline{z}$ is a combination of a code-word $\underline{c}$ and a symbol error pattern $\underline{e}'$, and since $\underline{c} \cdot \underline{w} = 0$ according to (1), then (5) can be written as follows:

$$A = (\underline{c} + \underline{e}').\underline{w}' = e_1' w_1' + e_2' w_2' + \cdots + e_n' \tag{6}$$

In the OSMLD codes, there are J check-sums orthogonal on each error symbol e_j', and this set of J check-sums can be used for estimating the error symbol e_j', with $1 \leq j \leq n$.

By removing the symbol on which the orthogonality holds, we obtain the following sum called the q-ary estimator on the symbol z_j:

$$B = A + z_n = z_1 w_1' + z_2 w_2' + \cdots + z_{n-1} w_{n-1}' \tag{7}$$

This time, the set of J orthogonal sum like (7) can be used to estimate the received symbol z_n. The Eqs. (6) and (7) are the non-binary adaptation of the two variants, namely A and B, given by Massey [7] as a generalization of the MLGD for the binary case. In the rest of the paper, the Eq. (7) is the one used to compute the parity checks.

Let the parity-check neighborhoods be the set $\mathcal{N}(i) = \{j : h_{i,j} \neq 0, 1 \leq j \leq n\}$ for $i = 1, \ldots, M$, where $h_{i,j}$ is the $(i,j)^{th}$ element of the parity check matrix H. Also, let the symbol neighborhoods be the set $\mathcal{M}(j) = \{i : h_{i,j} \neq 0, 1 \leq i \leq M\}$ for $j = 1, \ldots, n$.

Finally, denote channel reliability L, the vector $\underline{r}$ of magnitudes of the hard decision symbols z_j of the hard decision vector $\underline{z}$, the extrinsic information E, and the overall reliability R. The channel reliability L, the extrinsic information E and the overall reliability R, are all $q \times n$ matrices that represent reliabilities of each symbol $\hat{c}_j$ being one of the q possible symbols of $GF(q)$.

2.2   One-Step Majority-Logic Decodable Codes

The OSMLD codes are codes that can be decoded by taking a majority logic vote from a set of check equations orthogonal on each data symbol. When the code is cyclic, we need just one set of J equations orthogonal on the last symbol, then we cycle the data symbols so that the orthogonality becomes on the $(n-1)^{th}$ position, and so on until estimating all the n data symbols.

In the following, we show that certain properly formed check sums can be used for estimating the received symbols in z_j.

Suppose that there exist J vectors in the dual code $C^{\perp}$,

$$
\begin{aligned}
\underline{w_1'} &= (w_{1,1}', w_{1,2}', \ldots, w_{1,n}') \\
\underline{w_2'} &= (w_{2,1}', w_{2,2}', \ldots, w_{2,n}') \\
&\;\;\vdots \\
\underline{w_J'} &= (w_{J,1}', w_{J,2}', \ldots, w_{J,n}')
\end{aligned}
\tag{8}
$$

which have the following properties:

1. The nth component of each vector is " $\neq 0$ ", that is , $w_{1,n}' = w_{2,n}' = \ldots = w_{J,n}' = 1$.

2. For $i \neq n$, there is *at most* one vector whose i^{th} component is " $\neq 0$ "; for example, if $w_{1,i}' \neq 0$, then $w_{2,i}' = w_{3,i}' = \cdots = w_{J,i}' = 0$.

These J vectors are said to be orthogonal on the nth symbol position. We call them *orthogonal vectors*.

Now let us form J check sums from the set of orthogonal vectors in (8) using the variant B like in (7). We can form the J q-ary estimators on the symbol z_j in the following manner:

$$
\begin{aligned}
B_1 &= z_1 w_{1,1}' + z_2 w_{1,2}' + \cdots + z_{n-1} w_{1,n-1}' \\
B_2 &= z_1 w_{2,1}' + z_2 w_{2,2}' + \cdots + z_{n-1} w_{2,n-1}' \\
&\;\;\vdots \\
B_J &= z_1 w_{J,1}' + z_2 w_{J,2}' + \cdots + z_{n-1} w_{J,n-1}'
\end{aligned}
\tag{9}
$$

Therefore we see that the symbol z_n is checked by all the check sums above. Because of the second property of the J orthogonal vectors, any received symbol other than z_n is checked by at most one check sum. These J check sums are said to be *orthogonal on the symbol z_n* and they are crucial for the MLGD decoding algorithms. For more explanations, the reader is referred to [6] where the MLGD algorithms were well investigated.

Let d_{min} be the minimum distance of the code. Clearly, the majority-logic decoding is more effective for codes when the theoretical majority-logic error correction capability $t_{ML} = J/2$ is equal to or close to the error correcting capability $t = (d_{min} - 1)/2$ of the code; in other words, J should be equal to or close to $d_{min} - 1$. For the cyclic Euclidean geometry codes $d_{min} = J + 1$, therefore this family of codes are best suited for MLGD algorithms, and the bigger the set of equations is, the more powerful the decoding performance will be.

3   The Proposed Threshold Decoding Algorithm

The most commonly used metrics for soft-decision decoding, are the likelihood function, Euclidean distance, correlation, and correlation discrepancy. Hereafter, we give a description and a summary of the proposed Threshold Decoding algorithm.

Now suppose that the code-word $\underline{c}$ is modulated into $\underline{x}$, then transmitted in an AWGN channel, assuming BPSK modulation, as described in the previous section. Let $\underline{y} = (y_1, y_2, \ldots, y_n)$ be the received vector, the reliability of the received symbols being symbols from $GF(q)$ can be computed as by computing the log-likelihood ratios LLR's, as in [7].

$$LLR_j(\alpha_s) = ln\frac{P(c_j = \alpha_s|y_j)}{P(c_j = 0|y_j)} \tag{10}$$

Hence, there are q candidate values for each transmitted symbol y_j being $\alpha_s \in GF(q)$, $s = 0, 1, \ldots, q-1$. The most likely element is determined by comparing all LLR_j values and choosing the LLR with the largest magnitude, (i.e. the highest reliability).

However, in the non-binary case, this expression (10) only gives us the reliability of the nonzero element with respect to zero, and not to each other. Including these extra LLR's in the decoding procedure will provide more information about the reliability of each symbol, but the number of extra LLR's will increase exponentially with increasing alphabet size q.

By developing the conditional probability in (10),

$$P(y_j|x_j = \alpha'_s) = \frac{1}{\sqrt{\pi N_0}} \exp\left(-\sum_{b=0}^{m-1}(y_{j,b} - \alpha'_{s,b})^2\right) \tag{11}$$

a sum inside the computation reveals itself, which is simply the squared Euclidean distance between the received sequence y_j and the signal sequence of the constellation symbol $\alpha'_s = (-1)^{\alpha_s}$. We usually denote the this distance with $d_E^2(y_j, \alpha'_s)$ and it is equal to the sum

$$d_E^2(y_j, \alpha'_s) = \sum_{b=0}^{m-1}(y_{j,b} - \alpha'_{s,b})^2 \tag{12}$$

Consequently, maximizing $P(c_j = \alpha_s|y_j)$ or $\log P(y_j|x_j = \alpha'_s)$ is equivalent to minimizing the squared Euclidean distance $d_E^2(y_j, \alpha'_s)$. Therefore, a channel reliability can be carried out with the squared Euclidean distance as metric as follows: the received sequence y_j demodulated into the symbol α_s for which the squared Euclidean distance $d_E^2(y_j, \alpha'_s)$ is minimized.

Now, by expanding the right-hand side of (12), we obtain

$$d_E^2(y_j, \alpha'_s) = \sum_{b=0}^{m-1} y_{j,b}^2 + m - 2\sum_{b=0}^{m-1} y_{j,b} \cdot \alpha'_{s,b} \tag{13}$$

Algorithm 1. Summary of the proposed TD algorithm

Input: Received signal $\underline{y}$
Output: Non-binary Decoded sequence $\underline{\hat{c}}$
1: /* - - - - - - - - - - - Initialization - - - - - - - - - - - - */
2: **for** $j = (1:n)$ **do**
3:      **for** $s = (0:q-1)$ **do**
4:          $Lj\left(\alpha_s\right) = \sum_{b=0}^{m-1} y_{j,b} \cdot \alpha'_{s,b}$ / by using (14)
5:      **end for**
6:      $r_j = \max_s L_j\left(\alpha_s\right)$ / by using (15)
7:      $z_j = \arg_{\alpha_s} \max_s L_j\left(\alpha_s\right)$ / by using (16)
8: **end for**
9: $s_i = \sum_{p \in \mathcal{N}(i)} z_p h_{i,p}$ / by using (17)
10: **if** $\underline{s} == \underline{0}$ **then** return $\underline{z}$ as codeword
11: **else**
12:      /* - - - - - - - - - - - Processing - - - - - - - - - - - - */
13:      **for** $j = (n:1)$ **do**
14:          **for** $i = (0:J)$ **do**
15:              $B_i = \sum_{p \in \mathcal{N}'(i)} z_p w'_{i,p}$ / by using (18)
16:              $\omega_i = \min_{p \in \mathcal{N}'(i)} r_p$ / by using (20)
17:              $E_j\left(B_i\right) = E_j\left(B_i\right) + \omega_i$ / by using (21)
18:          **end for**
19:          $R_j = L_j + E_j$ / by using (22)
20:          $\hat{c}_j = \arg_{\alpha_s} \max_s R_j\left(\alpha_s\right)$ / by using (23)
21:          cyclically shift the vectors z and r
22:      **end for**
23: **end if**

In computing $d_E^2\left(y_j, \alpha'_s\right)$ for all the symbols in $GF\left(q\right)$, we see that $\sum_{b=0}^{m-1} y_{j,b}^2$ is a common term, and m is a constant. From (13) we readily see that minimizing $d_E^2\left(y_j, \alpha'_s\right)$ is equivalent to maximizing

$$L_j\left(\alpha^s\right) = \lambda\left(y_j, \alpha'_s\right) = \sum_{b=0}^{m-1} y_{j,b} \cdot \alpha'_{s,b} \tag{14}$$

The sum $\lambda\left(y_j, \alpha'_s\right)$ is called the correlation between the received sequence y_j and the signal sequence of the constellation symbol α'_s. Therefore, a channel reliability $L_j\left(\alpha^s\right)$ can be carried out with the correlation between the received sequence y_j and the signal sequence of the constellation symbol α'_s as a metric, where y_j is most likely the symbol α_s for which $L_j\left(\alpha^s\right)$ is maximized.

In this work, we adopt the approach of the correlation as a metric for computing channel reliability, because it is less complex than computing channel reliability based on log-likelihood ratios (10), and our goal is to produce a low complexity decoder.

Therefore we compute q values of $L_j\left(\alpha_s\right)$ for each received symbol y_j being $\alpha_s \in GF\left(q\right)$.

Let the magnitude of the reliability of the hard decision symbol be the maximum value:

$$r_j = \max_s L_j(\alpha_s) \tag{15}$$

Let $\underline{z} = (z_1, z_2, \ldots, z_n)$ be the hard decision vector of the received vector $\underline{y}$, computed from $L_j(\alpha_s)$. The symbol z_j is decided to be α_s for which $L_j(\alpha_s)$ is maximized.

$$z_j = \arg_{\alpha_s} \max_s L_j(\alpha_s) \tag{16}$$

Now we have the hard decision vector $\underline{z}$, we compute the syndrome to check if $\underline{z}$ is a code-word by computing $\underline{s} = \underline{z}.H^T$. We can also compute the syndrome using the parity check neighborhoods $\mathcal{N}(i)$ for $i = 1, \ldots, J$ as follows:

$$s_i = \sum_{p \in \mathcal{N}(i)} z_p h_{i,p} \tag{17}$$

and if $\underline{s} = \underline{0}$ then $\underline{z}$ is a codeword in C, and we stop the decoding. Otherwise, we continue the process by computing the q-ary estimator B_i.

Let the q-ary estimator neighborhoods $\mathcal{N}'(i) = \{j : w_{i,j} \neq 0, 1 \leq j \leq n-1\}$ for $i = 1, \ldots, J$, be the set of indices of non-zero coefficients of $\underline{w_i}$ involved in the q-ary estimator B_i, then B_i can be written:

$$B_i = \sum_{p \in \mathcal{N}'(i)} z_p w'_{i,p} \tag{18}$$

To compute the extrinsic reliability, we first have to compute the weighting coefficient of the ith equation as follows:

$$\omega_i = \log \left(\frac{1 + \prod_{p \in N'(i)} \tanh\left(\frac{r_p}{2}\right)}{1 - \prod_{p \in N'(i)} \tanh\left(\frac{r_p}{2}\right)} \right) \tag{19}$$

This formulation is too complex to be practical for our decoder, instead, we will use the simplification found in the literature [9], to compute the weighting factors of the J q-ary estimators on the symbol z_j:

$$\omega_i = \min_{p \in \mathcal{N}'(i)} r_p \tag{20}$$

This simplification leads to similar results, without performance degradation. Therefore, we will use it in this work.

Let $E_j(B_i)$ be the extrinsic reliability of the jth received symbol being B_i. The extrinsic reliability E is initialized by a $q \times n$ zero matrix, then it is computed as follows:

$$E_j(B_i) = E_j(B_i) + \omega_i \tag{21}$$

where for $i = 1, \ldots, J$, it adds the weight ω_i of the ith q-ary estimator to the location of the estimated value B_i, for each position j.

If the code is cyclic, the hard decision vector $\underline{z}$ and the vector of magnitudes $\underline{r}$ are cycled n times to compute R for every position j.

Then the overall reliability R_j for the jth position is computed by adding the channel reliability and the extrinsic information of that position.

$$R_j = L_j + E_j \tag{22}$$

Finally, the decision rule for the decoder is given by the following: $\underline{\hat{c}}$

$$\hat{c}_j = \arg_{\alpha_s} \max_s R_j(\alpha_s) \tag{23}$$

4   Results and Performances Analysis

4.1   Error Rates

Assume a transmission over the AWGN channel with BPSK modulation. Here, we exhibit the performance of our decoder in terms of Bit Error Rate (BER), Symbol Error Rate (SER), and Block Error Rate (BLER) of some EG codes present in Table 1, along with simulation settings.

Table 1. Simulation settings.

Channel codes	(n,k,d_{min})	Alphabet $GF(2^m)$
	EG(255,175,17)	$GF(256)$
	EG(1023,781,33)	$GF(32)$
	EG(4095,3367,65)	$GF(64)$
	RS(255,175,81)	$GF(256)$
Modulation	BPSK	
Channel	AWGN	
Decoding algorithms	MLGD, TD	
	HD-BM, ASD KV	
Simulation method	Monte Carlo	
Minimum number of transmitted blocks	1000	
Minimum number of residual symbol errors	200	

Figure 1 shows that decoding the EG(1023,781,33) over GF(32) with the proposed TD algorithm offers nearly 1 dB coding gain over the decoding it with the MLGD algorithm.

Figure 2 shows the block-error performances of the (255,175,17) EG code over GF(256) transmitted over the AWGN channel with BPSK modulation, decoded using the MLGD algorithm and using the TD algorithm. For a binary transmission, each symbol is expanded into 8-bit byte. The symbol-to-binary expansion results in a (2040,1400) binary code. At the receiving end, the received digits are grouped back into symbols in GF(2^8) for decoding.

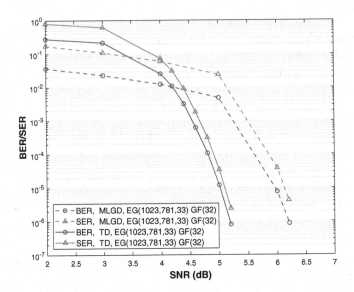

Fig. 1. Bit and symbol error rates of the (1023,781,33) EG cyclic code over GF(32) decoded with the TD algorithm compared with the MLGD.

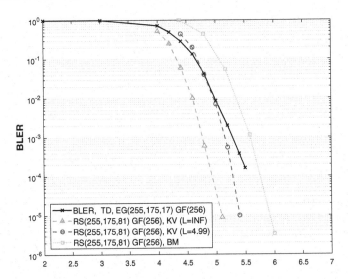

Fig. 2. BLER comparison of the proposed TD and MLGD algorithms with the HD-BM and ASD-KV algorithms.

Figure 2 also includes the block performances of the (255,175,81) RS code over $GF(2^8)$ decoded using the Hard-Decision Berlekamp-Massey (HD-BM), and using the Algebraic Soft Decoder Krotter-Vardy (ASD-KV) algorithm with interpolation complexity coefficients ∞ and 4.99. We notice that the performance of the TD falls between those of the HD-BM and the ASD-KV with interpolation complexity coefficients ∞. However, decoding the (255,175,81) RS code with algebraic decoding algorithms require computations in GF(256) to find the error-location polynomial and the error-value enumerator, whereas decoding the EG(255,175,17) code with MLGD or TD algorithms requires less complexity. As a result, decoding non-binary EG codes with threshold decoding can offer a trade-off to replace RS codes and their decoding. For more details about the complexity of the RS decoding algorithms, readers are referred to [6].

In Fig. 3 the symbol error rates of three cyclic codes are presented, namely the EG(1023,781,33) over GF(16), EG(1023,781,33) over GF(32) and EG(4095,3367,65) over GF(64) decoded with the proposed TD algorithm.

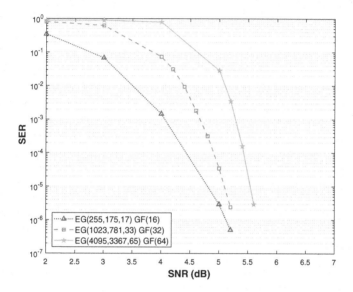

Fig. 3. Symbol error rates of three cyclic EG codes decoded with the TD algorithm.

4.2   Computational Complexity of the Threshold Decoder

In this section, we analyze the computational complexity of the proposed decoder based on the number of integer additions IAs, integer multiplications IMs, integer comparisons ICs, real additions RAs, real multiplications RMs, real comparisons RCs, field additions FAs and field multiplications FMs.

For the TD algorithm, let us consider the initialization step first. There are qn $L_j(\alpha_s)$s, to compute each one nmq RMs and $n(m-1)q$ RAs are needed. Then, to find r_j we need q RCs, and when r_j is found then z_j is straightforward. Therefore, to find the vectors $\underline{r}$ and $\underline{z}$, only qn RCs are needed. After that, the syndrome is computed and needs $n\rho$ FMs and $n(\rho-1)$ FAs.

For the rest of the decoding process, we need to compute the q-ary estimators which require $Jn(\rho-1)$ FMs and $Jn(\rho-2)$ FAs. Then, to compute ω_i's we need

Table 2. Computational complexity of the threshold decoder.

		FA	FM	RM	RA	RC
TD	Initialisation	$n\rho$	$n(\rho-1)$	nmq	$n(m-1)q$	$Jn(\rho-2)$
	Processing	$Jn(\rho-1)$	$Jn(\rho-2)$		$2Jn$	$n(q-1)$

$Jn(\rho-2)$ RCs, because there are $Jn\ \omega_i$ and each one needs $(\rho-2)$ RCs. After that, to compute E we need Jn RAs. Finally, to update R_j's we need to add the entries of E to L, which results in at most Jn additions. The final decision to find $\hat{c}$ is made by carrying at most $(q-1)n$ RCs. Table 2 summarizes the operations needed for our decoder.

5   Conclusion

The non-binary Threshold Decoding algorithm is an amelioration of the Majority Logic Decoding algorithm. It enhances the performance of this latter by taking into consideration the channel output before computing the orthogonal equations, which assigns weights to each one of these equations, which leads to more accurate results. The most pertinent point is that this decoding algorithm is very low complexity compared to most well-known decoders for non-binary codes, and since the decoding complexity is a key challenge, our decoder seems to be a good proposition. The TD is suitable for OSMLD codes and it is keeping the low complexity feature of the Majority Logic Decoding Algorithm. In future work, we aim to derive an iterative version of the proposed algorithm, which promises to give a low complexity decoding algorithm for non-binary codes. This idea seems interesting.

References

1. Belkasmi, M., Lahmer, M., Benchrifa, M.: Iterative threshold decoding of parallel concatenated block codes. In: 4th International Symposium on Turbo Codes & Related Topics; 6th International ITG-Conference on Source and Channel Coding, pp. 1–4. VDE (2006)
2. Belkasmi, M., Lahmer, M., Ayoub, F.: Iterative threshold decoding of product codes constructed from majority logic decodable codes. In: 2006 2nd International Conference on Information & Communication Technologies, vol. 2, pp. 2376–2381. IEEE (2006)
3. Berlekamp, E.R.: Algebraic Coding Theory, vol. 8. McGraw-Hill, New York (1968)
4. Kou, Y., Lin, S., Fossorier, M.P.: Low-density parity-check codes based on finite geometries: a rediscovery and new results. IEEE Trans. Inf. theory 47(7), 2711–2736 (2001)
5. Lin, S.: Number of information symbols in polynomial codes. IEEE Trans. Inf. Theory 18(6), 785–794 (1972)
6. Lin, S., Costello, D.: Error Control Coding. Prentice Hall, Upper Saddle River (2004)

7. Massey, J.L.: Threshold Decoding. MIT Press, Cambridge (1963)
8. Peterson, W.W., Weldon, E.J.: Error-Correcting Codes. MIT Press, Cambridge (1972)
9. Ryan, W., Lin, S.: Channel Codes: Classical and Modern. Cambridge University Press, Cambridge (2009)
10. Song, H., Cruz, J.R.: Reduced-complexity decoding of Q-ary LDPC codes for magnetic recording. IEEE Trans. Mag. **39**(2), 1081–1087 (2003)
11. Tang, H., Xu, J., Kou, Y., Lin, S., Abdel-Ghaffar, K.: On algebraic construction of Gallager and circulant low-density parity-check codes. IEEE Trans. Inf. Theory **50**(6), 1269–1279 (2004)
12. Tang, H., Xu, J., Lin, S., Abdel-Ghaffar, K.A.: Codes on finite geometries. IEEE Trans. Inf. Theory **51**(2), 572–596 (2005)
13. Weldon, E.J.: Euclidean geometry cyclic codes (No. Scientific-13). Hawaii University Honolulu Department of Electrical Engineering (1967)
14. Zeng, L., Lan, L., Tai, Y.Y., Zhou, B., Lin, S., Abdel-Ghaffar, K.A.: Construction of nonbinary cyclic, quasi-cyclic and regular LDPC codes: a finite geometry approach. IEEE Trans. Commun. **56**(3), 378–387 (2008)
15. Reed, I.: A class of multiple-error-correcting codes and the decoding scheme. Trans. IRE Prof. Group Inf. Theory **4**(4), 38–49 (1954)
16. Rudolph, L.: Geometric configuration and majority logic decodable codes. M.E.E. thesis, University of Oklahoma, Norman (1964)
17. Davey, M.C., MacKay, D.J.: Low density parity check codes over GF(q). In: 1998 Information Theory Workshop (Cat. No. 98EX131), pp. 70–71. IEEE (1998)
18. M'rabet, Z., Ayoub, F., Belkasmi, M., Yatribi, A., El Abidine, A.I.Z.: Non-binary Euclidean geometry codes: majority logic decoding. In: 2016 International Conference on Advanced Communication Systems and Information Security (ACOSIS), pp. 1–7. IEEE (2016)

Construction of Error Correcting Codes

On the Hamming and Symbol-Pair Distance of Constacyclic Codes of Length p^s over $\mathbb{F}_{p^m} + u\mathbb{F}_{p^m}$

Jamal Laaouine[✉]

Laboratory of Algebra, Geometry and Arithmetic, Department of Mathematics, Faculty of Sciences Dhar El Mahraz, University Sidi Mohamed Ben Abdellah, B.P. 1796 Fès-Atlas, Morocco
jamal.laaouine@usmba.ac.ma

Abstract. Let $\mathcal{R} = \mathbb{F}_{p^m} + u\mathbb{F}_{p^m}, u^2 = 0$, be the finite commutative chain ring with unity, where p is a prime number, m is a positive integer and $\mathbb{F}_{p^m}$ is the finite field with p^m elements. In this work, we give a simple and short proof of classification of all γ-constacyclic codes of length p^s over $\mathcal{R}$, that is, ideals of the ring $\mathcal{R}[x]/\langle x^{p^s} - \gamma \rangle$, where γ is a nonzero element of the field $\mathbb{F}_{p^m}$. This allows us to study the Hamming and symbol-pair distance distributions of all such codes.

Keywords: Constacyclic codes · Hamming distance · Symbol-pair distance · Finite chain rings

1 Introduction

Constacyclic codes are generalizations of cyclic and negacyclic codes, and they play a significant role in coding theory. This family of codes can be efficiently encoded and decoded using a simple shift register, which explains their preferred role in engineering. For both theoretical and practical reasons, constacyclic codes received a great deal of attention from many researchers.

Given a unit γ of a finite ring R, γ-constacyclic codes of length n over R are in correspondence with ideals of the ring $R[x]/\langle x^n - \gamma \rangle$. The case when the code length n is divisible by the characteristic p of the residue field of the underlying ring yields the so-called repeated-root codes. Let p be a prime, s, m be positive integers, and let $\mathbb{F}_{p^m}$ be the finite field of order p^m. From [4,5], the algebraic structure and Hamming distances of constacyclic codes of length p^s over $\mathbb{F}_{p^m}$ were determined. Dinh [5] classified and gave all constacyclic codes of length p^s over the finite commutative chain ring $\mathbb{F}_{p^m}[u]/\langle u^2 \rangle$.

Symbol-pair codes desired to protect against a certain number of pair-errors were first introduced by Cassuto and Blaum [1,2]. They established a simple but important connection between the symbol-pair distance and the Hamming distance by [2, Theorem 2]. Using algebraic methods, Cassuto and Litsyn [3] constructed cyclic symbol-pair codes. In 2012, Yaakobi et al. [14] considered and

© Springer Nature Switzerland AG 2020
M. Belkasmi et al. (Eds.): ACOSIS 2019, CCIS 1264, pp. 137–154, 2020.
https://doi.org/10.1007/978-3-030-61143-9_12

gave a better lower-bound on the symbol-pair distances for binary cyclic codes. By using the definition of parity-check matrix for decoding symbol-pair codes, Hirotomo et al. [9] proposed a new algorithm that improved the algorithms of Cassuto et al. [3] and Yaakobi et al. [14] for decoding symbol-pair codes. Particularly, in 2015, by extending [3, Theorem 10], Kai et al. [11] provided a new lower bound on simple-root constacyclic codes. In another related work, Dinh et al. [8] determined symbol-pair distances of all constacyclic codes of length p^s over $\mathbb{F}_{p^m}$.

Motivated by these investigations, we completely solve the problem of determination of the Hamming and symbol-pair distance of γ-constacyclic codes of length p^s over $\mathcal{R} = \mathbb{F}_{p^m} + u\mathbb{F}_{p^m}(u^2 = 0)$, in this paper, where $\gamma \in \mathbb{F}_{p^m}^*$. After giving some preliminaries in Sect. 2, the rest of the paper is organized as follows. In Sect. 3, we give another proof of a result cited in Dinh paper (see [5, 5.4, 6.1]). Section 4 addresses the Hamming distances of γ-constacyclic codes of length p^s over $\mathcal{R}$, where γ is a nonzero element of the field $\mathbb{F}_{p^m}$. Theorems 3.3, 3.4 in [7] and Theorem 5.2 (Type 3 and 4) in [12] are not completely true, so we had to add some assumptions to correct them completely. In Sect. 4, we continue to study symbol-pair distances of γ-constacyclic codes of length p^s over $\mathcal{R}$, Theorem 4.3 (Type 3 and 4) in [7] is not entirely true, so we had to add some hypotheses to correct it completely. Section 6 concludes the paper.

2   Some Preliminaries

All rings are commutative rings. A ring R is a principal ideal ring if its ideals are principal. R is called a local ring if R has a unique maximal ideal. Finally, R is called a chain ring if its ideals are linearly ordered under set theory inclusion.

The following result is well-known for the class of finite commutative chain rings (see [6]).

Proposition 1. *Let R be a finite ring, then the following conditions are equivalent:*

(i) R is a local ring and the maximal ideal M of R is principal,
(ii) R is a local principal ideal ring,
(iii) R is a chain ring.

A subset $C \subseteq R^n$ is called a linear code of length n over R if C is an R-submodule of R^n and the ring R is referred to as the alphabet of C. For a unit γ of R, the γ-constacyclic (γ-twisted) shift τ_γ on R^n is the shift

$$\tau_\gamma((x_0, x_1, \ldots, x_{n-1})) = (\gamma x_{n-1}, x_0, x_1, \ldots, x_{n-2}),$$

and a code C is said to be constacyclic, or specifically, γ-constacyclic if C is invariant under the γ-constacyclic shift τ_γ. In light of this definition, if $\gamma = 1$, those γ-constacyclic codes are called cyclic codes, and when $\gamma = -1$, such γ-constacyclic codes are called negacyclic codes.

Each codeword $c = (c_0, c_1, \ldots, c_{n-1}) \in C$ can be identified with its polynomial representation $c(x) = c_0 + c_1 x + \cdots + c_{n-1} x^{n-1}$. Then in the ring $R[x]/\langle x^n - \gamma\rangle$, $xc(x)$ corresponds to a γ-constacyclic shift of $c(x)$. From that, the following fact follows at once (cf. [10,13]).

Proposition 2. *A linear code C of length n is γ-constacyclic over R if and only if C is an ideal of $R[x]/\langle x^n - \gamma\rangle$.*

For a word $\boldsymbol{x} = (x_0, x_1, \ldots, x_{n-1}) \in R^n$, the Hamming weight of $\boldsymbol{x}$, denoted by $wt_H(\boldsymbol{x})$, is the number of nonzero entry x_i for $0 \leq i \leq n - 1$. The Hamming distance $d_H(\boldsymbol{x}, \boldsymbol{y})$ of two words $\boldsymbol{x}$ and $\boldsymbol{y}$ is the Hamming weight $wt_H(\boldsymbol{x} - \boldsymbol{y})$ of $\boldsymbol{x} - \boldsymbol{y}$. The minimum Hamming weight $wt_H(C)$ and the minimum Hamming distance $d_H(C)$ of a linear code C are the same, i.e.,

$$d_H(C) = \min\{wt_H(\boldsymbol{x}) \mid 0 \neq \boldsymbol{x} \in C\}.$$

For any nonzero $\gamma \in \mathbb{F}_{p^m}$, linear γ-constacyclic codes of p^s length over $\mathbb{F}_{p^m}$ are precisely the ideals in $\mathbb{F}_{p^m}[x]/\langle x^{p^s} - \gamma\rangle$. By applying the Division Algorithm, there exist nonnegative integers γ_q, γ_r such that $s = \gamma_q m + \gamma_r$ with $0 \leq \gamma_r \leq m - 1$. Let $\gamma_0 = \gamma^{p^{(\gamma_q+1)m-s}} = \gamma^{p^{m-\gamma_r}}$. Then $\gamma_0^{p^s} = \gamma^{p^{(\gamma_q+1)m}} = \gamma$.

In [4] and [5], the algebraic structure and Hamming distances of γ-constacyclic codes of length p^s over $\mathbb{F}_{p^m}$ were established and given by the following theorem.

Theorem 1 (cf. [4,5]). *Let C be a γ-constacyclic code of length p^s over $\mathbb{F}_{p^m}$. Then $C = \langle(x - \gamma_0)^i\rangle \subseteq \mathbb{F}_{p^m}[x]/\langle x^{p^s} - \gamma\rangle$, for $i \in \{0, 1, \ldots, p^s\}$, and its Hamming distance $d_H(C)$ is completely determined by:*

$$d_H(C) = \begin{cases} \bullet \; 1, & \text{if } i = 0, \\[2mm] \bullet \; (n+1)p^k, & \text{if} \\ \quad p^s - pr + (n-1)r + 1 \leq i \leq p^s - pr + nr, \\ \quad \text{where } r = p^{s-k-1},\; 1 \leq n \leq p - 1 \\ \quad \text{and } 0 \leq k \leq s - 1, \\[2mm] \bullet \; 0, & \text{if } i = p^s. \end{cases}$$

Let Σ be the code alphabet consisting of q elements, whose elements are called symbols. In symbol-pair read channels, a codeword $\boldsymbol{x} = (x_0, x_1, \ldots, x_{n-1})$ is represented as

$$\pi(\boldsymbol{x}) = ((x_0, x_1), (x_1, x_2), \ldots, (x_{n-1}, x_0)) \in (\Sigma^2)^n.$$

For any two symbol pairs (e, f) and (g, h), say $(e, f) = (g, h)$ if both $e = g$ and $f = h$. An important parameter of symbol-pair codes is the symbol-pair distance. Given $\boldsymbol{x} = (x_0, x_1, \ldots, x_{n-1})$, $\boldsymbol{y} = (y_0, y_1, \ldots, y_{n-1})$, the symbol-pair distance between $\boldsymbol{x}$ and $\boldsymbol{y}$ is given in [1] by using Hamming distance over the alphabet (Σ, Σ) as follows:

$$d_{sp}(\boldsymbol{x}, \boldsymbol{y}) = d_H(\pi(\boldsymbol{x}), \pi(\boldsymbol{y})) = |\{i \mid (x_i, x_{i+1}) \neq (y_i, y_{i+1})\}|.$$

The minimum pair distance of a code C is defined to be

$$d_{sp}(C) = \min\{d_{sp}(\boldsymbol{x}, \boldsymbol{y}) \mid \boldsymbol{x}, \boldsymbol{y} \in C, \boldsymbol{x} \neq \boldsymbol{y}\}.$$

The symbol-pair weight of a vector $\boldsymbol{x}$ is defined as the Hamming weight of its symbol-pair vector $\pi(\boldsymbol{x})$:

$$wt_{sp}(\boldsymbol{x}) = wt_H(\pi(\boldsymbol{x})) = |\{i \mid (x_i, x_{i+1}) \neq (0,0), 0 \leq i \leq n-1, x_n = x_0\}|.$$

In addition, if the code C is linear symbol-pair code, its symbol-pair weight and symbol-pair distance are coincided, i.e.,

$$d_{sp}(C) = \min\{wt_{sp}(\boldsymbol{x}) \mid 0 \neq \boldsymbol{x} \in C\}.$$

In [8], the symbol-pair distances of γ-constacyclic codes of length p^s over $\mathbb{F}_{p^m}$ were established and given by the following theorem.

Theorem 2 (cf. [8]). *Let $C = \langle (x - \gamma_0)^i \rangle, 0 \leq i \leq p^s$, be a γ-constacyclic code of length p^s over $\mathbb{F}_{p^m}$. Then the symbol-pair distance $d_{sp}(C)$ of C is completely determined by:*

$$d_{sp}(C) = \begin{cases} \bullet \ 2, \quad if \ i = 0, \\[2mm] \bullet \ 3p^k, \quad if \ i = p^s - p^{s-k} + 1, \\ \qquad where \ 0 \leq k \leq s-2, \\[2mm] \bullet \ 4p^k, \quad if \\ \quad p^s - p^{s-k} + 2 \leq i \leq p^s - p^{s-k} + p^{s-k-1}, \\ \qquad where \ 0 \leq k \leq s-2, \\[2mm] \bullet \ 2(\sigma+2)p^k, \quad if \\ \quad p^s - pr + \sigma r + 1 \leq i \leq p^s - pr + (\sigma+1)r, \\ \qquad where \ r = p^{s-k-1}, \ \ 0 \leq k \leq s-2 \\ \qquad and \ 1 \leq \sigma \leq p-2, \\[2mm] \bullet \ (\sigma+2)p^{s-1}, \quad if \ i = p^s - p + \sigma, \\ \qquad where \ 1 \leq \sigma \leq p-2, \\[2mm] \bullet \ p^s, \quad if \ i = p^s - 1, \\[2mm] \bullet \ 0, \quad if \ i = p^s. \end{cases}$$

The following lemma plays an important role in studying algebraic structures of γ-constacyclic code of length p^s over the ring $\mathcal{R}$.

Lemma 1. *Let A and B be two commutative rings. Let $\varphi : A \longrightarrow B$ be a subjective ring homomorphism. Assume that B is a principal ideal ring and $\ker(\varphi)$ is a principal ideal $\langle \pi \rangle = \pi A$. Let I be an ideal of the ring A. Then there exists an $a \in I$ such that*

$$I = aA + \pi J,$$

where $J = (I : \pi) = \{r \in A : r\pi \in I\}$.

Proof. Let $b \in B$ such that $\varphi(I) = Bb$. Let $a \in I$ such that $\varphi(a) = b$. If $x \in I$ then there exists an $w \in B$ such that $\varphi(x) = bw = \varphi(a)\varphi(v)$, where $v \in A$ and $\varphi(v) = w$. This implies that $y = x - av \in \ker(\varphi)$, which also implies that $x = av + y$ with $y \in \ker(\varphi) \cap I$. Hence, we get $I \subseteq Aa + \ker(\varphi) \cap I$, which means that $I = aA + \ker(\varphi) \cap I$. Since $\ker(\varphi) = \pi A$, we have $\ker(\varphi) \cap I = \pi J$, where $J = (I : \pi)$. Therefore, the desired result follows.   □

Let $\mathbb{F}_{p^m}$ be a finite field of p^m elements, where p is prime number, and denote

$$\mathcal{R} = \mathbb{F}_{p^m} + u\mathbb{F}_{p^m}(u^2 = 0).$$

The ring $\mathcal{R}$ can be expressed as $\mathcal{R} = \mathbb{F}_{p^m}[u]/\langle u^2 \rangle = \{a + bu \mid a, b \in \mathbb{F}_{p^m}\}$. It is easy to verify that $\mathcal{R}$ is a local ring with maximal ideal $\langle u \rangle = u\mathbb{F}_{p^m}$. Therefore, by Proposition 1, it is a chain ring. Every invertible element in $\mathcal{R}$ is of the form: $\alpha + \beta u$ where $\alpha, \beta \in \mathbb{F}_{p^m}$ and $\alpha \neq 0$.

3   Constacyclic Codes of Length p^s over $\mathcal{R}$

From now onwards, we shall focus our attention on γ-constacyclic codes of length p^s over $\mathcal{R}$, i.e., ideals of the ring

$$\mathcal{R}_\gamma = \mathcal{R}[x]/\langle x^{p^s} - \gamma \rangle,$$

where γ is a nonzero element of the field $\mathbb{F}_{p^m}$.

Lemma 2 (cf. [5]). *The followings hold in $\mathcal{R}_\gamma$:*

(i) $x - \gamma_0$ is nilpotent with the nilpotency index p^s.
(ii) Any $f(x) \in \mathcal{R}_\gamma$ can be written uniquely as:

$$f(x) = \sum_{j=0}^{p^s-1} a_j(x - \gamma_0)^j + u \sum_{j=0}^{p^s-1} b_j(x - \gamma_0)^j$$

$$= a_0 + (x - \gamma_0) \sum_{j=1}^{p^s-1} a_j(x - \gamma_0)^{j-1} + u \sum_{j=0}^{p^s-1} b_j(x - \gamma_0)^j,$$

where a_j and b_j are elements of $\mathbb{F}_{p^m}$. Furthermore, $f(x)$ is invertible if and only if $a_0 \neq 0$.

Proposition 3 (cf. [5]). *The ring $\mathcal{R}_\gamma = \mathcal{R}[x]/\langle x^{p^s} - \gamma \rangle$ is a local ring with the maximal ideal $\langle u, x - \gamma_0 \rangle$, but it is not a chain ring.*

In the following we give an alternative short proof of a deep result cited in Dinh paper (see [5, 5.4, 6.1]).

Theorem 3 cf. [5, 5.4, 6.1]). *The γ-constacyclic codes of length p^s over $\mathcal{R}$, i.e., ideals of the ring $\mathcal{R}_\gamma$, are*

Type 1 $(\mathcal{C}_1)$ *(trivial ideals):*

$$\langle 0 \rangle, \ \langle 1 \rangle.$$

Type 2 $(\mathcal{C}_2)$ *(principal ideals with nonmonic polynomial generators):*

$$\mathcal{C}_2 = \langle u(x - \gamma_0)^\tau \rangle, \quad where \ 0 \leq \tau \leq p^s - 1.$$

Type 3 $(\mathcal{C}_3)$ *(principal ideals with monic polynomial generators):*

$$\mathcal{C}_3 = \langle (x - \gamma_0)^\delta + u(x - \gamma_0)^t h(x) \rangle,$$

where $1 \leq \mathsf{T} \leq \delta \leq p^s - 1$, $0 \leq t < \mathsf{T}$, *either* $h(x)$ *is* 0 *or* $h(x)$ *is a unit in* $\mathcal{R}_\gamma$ *of the form* $\sum_{i=0}^{\mathsf{T}-t-1} h_i (x - \gamma_0)^i$ *with* $h_i \in \mathbb{F}_{p^m}$ *and* $h_0 \neq 0$. *Here* T *is the smallest integer satisfying* $u(x - \gamma_0)^\mathsf{T} \in \mathcal{C}_3$.

Type 4 $(\mathcal{C}_4)$ *(non principal ideals):*

$$\mathcal{C}_4 = \langle (x - \gamma_0)^\delta + u(x - \gamma_0)^t h(x), u(x - \gamma_0)^\omega \rangle,$$

where $0 \leq \omega < \mathsf{T} \leq \delta \leq p^s - 1$, $0 \leq t < \omega$, *either* $h(x)$ *is* 0 *or* $h(x)$ *is a unit in* $\mathcal{R}_\gamma$ *of the form* $\sum_{i=0}^{\omega-t-1} h_i (x - \gamma_0)^i$ *with* $h_i \in \mathbb{F}_{p^m}$, $h_0 \neq 0$ *and* T *is the smallest integer such that* $u(x - \gamma_0)^\mathsf{T} \in \mathcal{C}_3$.

Proof. Let μ be the map from $\mathcal{R}_\gamma$ to $\mathbb{F}_{p^m}[x]/\langle x^{p^s} - \gamma \rangle$ given by $\mu(f(x)) = f(x) \ mod \ u$. Clearly, μ is a subjective ring homomorphism. It is obvious to see that ideals of **Type 1**$(\mathcal{C}_1)$ are trivial ideals. Let I be an arbitrary nontrivial ideal of $\mathcal{R}_\gamma$. Now the following two cases arise: **Case 1** $: I \subseteq \langle u \rangle$ and **Case 2** $: I \nsubseteq \langle u \rangle$.

∗ **Case 1.** First suppose that $I \subseteq \langle u \rangle$. Let $J = \{f(x) \in \mathcal{R}_\gamma : f(x)u \in I\} = (I : u)$. Then, by Lemma 1, $I = uJ$. Here $\mu(J)$ is a nonzero ideal of the ring $\mathbb{F}_{p^m}[x]/\langle x^{p^s} - \gamma \rangle$. Then $\mu(J) = \langle (x - \gamma_0)^\tau \rangle$ for some τ, $0 \leq \tau \leq p^s - 1$. By Lemma 1, there exists $f(x) \in \mathcal{R}_\gamma$, such that $J = \langle (x - \gamma_0)^\tau + uf(x) \rangle$. Then

$$I = \langle u(x - \gamma_0)^\tau \rangle,$$

which is of **Type 2** $(\mathcal{C}_2)$.

∗ **Case 2.** Next suppose that $I \nsubseteq \langle u \rangle$. Then $\mu(I)$ is a nonzero ideal of the ring $\mathbb{F}_{p^m}[x]/\langle x^{p^s} - \gamma \rangle$. Hence there is an integer $\delta \in \{0, 1, \ldots, p^s - 1\}$ such that $\mu(I) = \langle (x - \gamma_0)^\delta \rangle$. Next by lemma 1, there exists $f(x) \in \mathcal{R}_\gamma$, such that $I = \langle (x - \gamma_0)^\delta + uf(x) \rangle + uJ$, where $J = (I : u)$. We have $uJ \subseteq \langle u \rangle$ and we see that $(uJ : u) = (\langle u \rangle \cap I : u) = (\langle u \rangle : u) \cap (I : u) = (I : u) = J \nsubseteq \langle u \rangle$ (because $I \subseteq J$). This, by **Case 1**, implies that

$$uJ = \langle u(x - \gamma_0)^\omega \rangle,$$

where $0 \leq \omega \leq \delta \leq p^s - 1$. From this, we obtain

$$I = \langle (x - \gamma_0)^\delta + uf(x), u(x - \gamma_0)^\omega \rangle.$$

Let us write

$$f(x) = \sum_{j=0}^{p^s-1} a_j(x - \gamma_0)^j + u \sum_{j=0}^{p^s-1} b_j(x - \gamma_0)^j,$$

where a_j and b_j are elements of $\mathbb{F}_{p^m}$, then

$$(x - \gamma_0)^\delta + uf(x) = (x - \gamma_0)^\delta + u \sum_{j=0}^{\omega-1} a_j(x - \gamma_0)^j$$

$$+ u(x - \gamma_0)^\omega \sum_{j=0}^{p^s-\omega-1} a_{j+\omega}(x - \gamma_0)^j.$$

Since $(x - \gamma_0)^\delta + uf(x) \in I$ and $u(x - \gamma_0)^\omega \in I$, it follows that $(x - \gamma_0)^\delta + u\sum_{j=0}^{\omega-1} a_j(x - \gamma_0)^j \in I$. From this, it follows that

$$I = \left\langle (x - \gamma_0)^\delta + u \sum_{j=0}^{\omega-1} a_j(x - \gamma_0)^j, u(x - \gamma_0)^\omega \right\rangle$$

$$= \langle (x - \gamma_0)^\delta + u(x - \gamma_0)^t h(x), u(x - \gamma_0)^\omega \rangle,$$

where $0 \leq t < \omega$ and either $h(x)$ is 0 or $h(x)$ is a unit in $\mathcal{R}_\gamma$ of the form $\sum_{j=0}^{\omega-t-1} a_j(x - \gamma_0)^j$ with $a_j \in \mathbb{F}_{p^m}$, $a_0 \neq 0$.

Let T be the smallest integer such that $u(x - \gamma_0)^{\mathsf{T}} \in \langle (x - \gamma_0)^\delta + u(x - \gamma_0)^t h(x) \rangle$. Since $u(x - \gamma_0)^\delta \in \langle (x - \gamma_0)^\delta + u(x - \gamma_0)^t h(x) \rangle$, we get $\mathsf{T} \leq \delta$. As $u(x - \gamma_0)^{\mathsf{T}} \in \langle u(x - \gamma_0)^\omega \rangle$ with the reason that $(x - \gamma_0)^{\mathsf{T}} \in J := (I : u)$, we must have $\omega \leq \mathsf{T}$. We now divide into two subcases.

○ Case 2a. If $\omega = \mathsf{T}$, then

$$I = \langle (x - \gamma_0)^\delta + u(x - \gamma_0)^t h(x) \rangle,$$

which are ideals of Type 3 ($\mathcal{C}_3$).

○ Case 2b. If $\omega < \mathsf{T}$, then

$$I = \langle (x - \gamma_0)^\delta + u(x - \gamma_0)^t h(x), u(x - \gamma_0)^\omega \rangle,$$

which means that I is of Type 4 ($\mathcal{C}_4$).

This completes the proof of the theorem.

□

Proposition 4 (cf. [5]). *Let* T *be the smallest integer such that* $u(x - \gamma_0)^{\mathsf{T}} \in$ $\mathcal{C}_3 = \langle (x - \gamma_0)^\delta + u(x - \gamma_0)^t h(x) \rangle$. *Then*

$$\mathsf{T} = \begin{cases} \delta, & if\, h(x) = 0, \\ \min\{\delta, p^s - \delta + t\}, & if\, h(x) \neq 0. \end{cases}$$

4   Hamming Distance

As we mentioned in Sect. 3 the γ-constacyclic codes of length p^s over $\mathcal{R}$ are precisely the ideals of the ring $\mathcal{R}_\gamma$. In order to compute the Hamming distances of all γ-constacyclic codes of length p^s over $\mathcal{R}$, we count the Hamming distance of the ideals of the ring $\mathcal{R}_\gamma$ as classified into 4 types in Theorem 3. It is easy to see that $d_H(\mathcal{C}) = 0$ when $\mathcal{C} = \{0\}$, and that $d_H(\mathcal{C}) = 1$ when $\mathcal{C} = \{1\}$. The Hamming distances of γ-constacyclic codes of **Type 2** were determined in [7].

In order to compute the Hamming distances of those codes for the rest cases (**Type 3** and **4**), we first observe that

$$wt_H(a(x) + ub(x)) \geq wt_H(a(x)), \tag{1}$$

where $a(x), b(x) \in \mathbb{F}_{p^m}[x]/\langle x^{p^s} - \gamma \rangle$.

Note that $\mathbb{F}_{p^m}$ is a subring of $\mathcal{R}$, for a code $\mathcal{C}$ over $\mathcal{R}$, we denote $d_H(\mathcal{C}_{\mathbb{F}})$ as the Hamming distance of $\mathcal{C}|_{\mathbb{F}_{p^m}}$.

In [7, Theorem 3.3], Dinh et al. stated that: $d_H(\mathcal{C}_3) = d_H(\langle ((x-\gamma_0)^\delta)_{\mathbb{F}} \rangle)$. Unfortunately, this result is not true in general, which we illustrate in the following example.

Example 1. Let $\mathcal{R} = \mathbb{F}_{7^m} + u\mathbb{F}_{7^m}$, where $u^2 = 0$. Consider the γ-constacyclic code $\mathcal{C}_3 = \langle (x - \gamma_0)^5 + u(x - \gamma_0)h(x) \rangle$ of length 7 over $\mathcal{R}$, where $h(x) \neq 0$. Here $p = 7$, $s = 1$, $\delta = 5$ and $t = 1$. Then $\mathsf{T} = 3$, which implies that $\langle u(x - \gamma_0)^3 \rangle \subseteq \mathcal{C}_3$. This implies that $d_H(\langle u(x - \gamma_0)^3 \rangle) = d_H(\langle ((x - \gamma_0)^3)_{\mathbb{F}} \rangle) \geq d_H(\mathcal{C}_3)$. By Theorem 1, we see that $d_H(\langle ((x - \gamma_0)^3)_{\mathbb{F}} \rangle) = 4$ and $d_H(\langle ((x - \gamma_0)^5)_{\mathbb{F}} \rangle) = 6$. Now, we observe that $d_H(\mathcal{C}_3) \neq d_H(\langle ((x - \gamma_0)^5)_{\mathbb{F}} \rangle)$. This shows that there is an error in Theorem 3.3 of Dinh et al. [7].

In the following theorem, we shall rectify the error in Theorem 3.3 of Dinh et al. [7].

Theorem 4. *Let* $\mathcal{C}_3 = \langle (x - \gamma_0)^\delta + u(x - \gamma_0)^t h(x) \rangle$ *be a* γ-*constacyclic codes of length* p^s *over* $\mathcal{R}$ *of* **Type 3** *(as classified in Theorem 3), where* $1 \leq \mathsf{T} \leq \delta \leq p^s - 1$, $0 \leq t < \mathsf{T}$, *either* $h(x)$ *is 0 or* $h(x)$ *is a unit and* T *is the smallest integer satisfying* $u(x - \gamma_0)^{\mathsf{T}} \in \mathcal{C}_3$, *i.e.,*

$$\mathsf{T} = \begin{cases} \delta, & if\, h(x) = 0, \\ \min\{\delta, p^s - \delta + t\}, & if\, h(x) \neq 0. \end{cases}$$

Then the Hamming distance of $\mathcal{C}_3$ *is*

$$d_H(\mathcal{C}_3) = (n + 1)p^k,$$

where $p^s - pr + (n-1)r + 1 \leq \mathsf{T} \leq p^s - pr + nr$, $r = p^{s-k-1}$, $1 \leq n \leq p-1$ *and* $0 \leq k \leq s - 1$.

Proof. Let $\mathcal{C}_3 = \langle (x - \gamma_0)^\delta + u(x - \gamma_0)^t h(x) \rangle$ be of Type 3. And let $c(x)$ be an arbitrary nonzero element of $\mathcal{C}_3$, then $c(x)$ can be written as

$$c(x) = [f(x) + ug(x)][(x - \gamma_0)^\delta + u(x - \gamma_0)^t h(x)],$$

where $f(x), g(x) \in \mathbb{F}_{p^m}[x]$. We consider two cases.

* Case 1: $f(x) = 0$. In this case, $g(x) \neq 0$. We have
$c(x) = ug(x)(x - \gamma_0)^\delta$, which implies that

$$\begin{aligned}
wt_H(c(x)) &= wt_H(ug(x)(x - \gamma_0)^\delta) \\
&\geq d_H(\langle u(x - \gamma_0)^\delta \rangle) = d_H(\langle (x - \gamma_0)^\delta \rangle_\mathbb{F}).
\end{aligned}$$

Since, $\langle (x - \gamma_0)^\delta \rangle \subseteq \langle (x - \gamma_0)^\mathsf{T} \rangle$, we have

$$d_H(\langle (x - \gamma_0)^\delta \rangle_\mathbb{F}) \geq d_H(\langle (x - \gamma_0)^\mathsf{T} \rangle_\mathbb{F}).$$

* Case 2: $f(x) \neq 0$. Then we have

$$c(x) = f(x)(x - \gamma_0)^\delta + u[g(x)(x - \gamma_0)^\delta + f(x)(x - \gamma_0)^t h(x)].$$

By (1), we obtain that

$$\begin{aligned}
wt_H(c(x)) &\geq wt_H(f(x)(x - \gamma_0)^\delta) \\
&\geq d_H(\langle (x - \gamma_0)^\delta \rangle_\mathbb{F}) \\
&\geq d_H(\langle (x - \gamma_0)^\mathsf{T} \rangle_\mathbb{F}).
\end{aligned}$$

From this, we get $wt_H(c(x)) \geq d_H(\langle (x - \gamma_0)^\mathsf{T} \rangle_\mathbb{F})$ for each $c(x)$ nonzero element of $\mathcal{C}_3$. This implies that

$$d_H(\mathcal{C}_3) \geq d_H(\langle (x - \gamma_0)^\mathsf{T} \rangle_\mathbb{F}) \tag{2}$$

On the other hand we have that

$$\langle u(x - \gamma_0)^\mathsf{T} \rangle \subseteq \mathcal{C}_3,$$

which implies that

$$d_H(\langle (x - \gamma_0)^\mathsf{T} \rangle_\mathbb{F}) = d_H(\langle u(x - \gamma_0)^\mathsf{T} \rangle) \geq d_H(\mathcal{C}_3) \tag{3}$$

Now by (2) and (3), we get

$$d_H(\mathcal{C}_3) = d_H(\langle (x - \gamma_0)^\mathsf{T} \rangle_\mathbb{F}).$$

The rest of the proof follows from Theorem 1 and the discussion above. $\square$

Corollary 1. *Following the same notations as in Theorem 4, we have the following:*

1. *If $h(x)$ is 0 or $h(x)$ is a unit and $1 \leq \delta \leq \frac{p^s+t}{2}$, then*

$$d_H(\mathcal{C}_3) = d_H(\langle (x - \gamma_0)^\delta \rangle_{\mathbb{F}})$$
$$= (n+1)p^k,$$

where $p^s - pr + (n-1)r + 1 \leq \delta \leq p^s - pr + nr$, $r = p^{s-k-1}$, $1 \leq n \leq p - 1$ and $0 \leq k \leq s - 1$.

2. *If $h(x)$ is a unit and $\frac{p^s+t}{2} < \delta \leq p^s - 1$, then*

$$d_H(\mathcal{C}_3) = d_H(\langle (x - \gamma_0)^{p^s-\delta+t} \rangle_{\mathbb{F}})$$
$$= (n+1)p^k,$$

where $t + pr - nr \leq \delta \leq t + pr - (n-1)r - 1$, $r = p^{s-k-1}$, $1 \leq n \leq p - 1$ and $0 \leq k \leq s - 1$.

Proof. When $h(x) = 0$ or $h(x) \neq 0$ and $1 \leq \delta \leq \frac{p^s+t}{2}$, we have $\mathsf{T} = \delta$ and when $h(x) \neq 0$ and $\frac{p^s+t}{2} < \delta \leq p^s - 1$, then $\mathsf{T} = p^s - \delta + t$. The rest of the proof follows from Theorem 4. □

Theorem 5. *Let $\mathcal{C}_4 = \langle (x-\gamma_0)^\delta + u(x-\gamma_0)^t h(x), u(x-\gamma_0)^\omega \rangle$ be a γ-constacyclic codes of length p^s over $\mathcal{R}$ of Type 4 (as determined in Theorem 3), where $0 \leq \omega < \delta \leq p^s - 1$ and either $h(x)$ is 0 or $h(x)$ is a unit. Then the Hamming distance of $\mathcal{C}_4$ is given by*

$$d_H(\mathcal{C}_4) = \begin{cases} \bullet \ 1, \ if \ \omega = 0, \\[2mm] \bullet \ (n+1)p^k, \ if \\ \quad p^s - pr + (n-1)r + 1 \leq \omega \leq p^s - pr + nr, \\ \qquad where \ r = p^{s-k-1}, \ 1 \leq n \leq p - 1 \\ \qquad\qquad and \ 0 \leq k \leq s - 1. \end{cases}$$

Proof. Let $\mathcal{C}_4 = \langle (x - \gamma_0)^\delta + u(x - \gamma_0)^t h(x), u(x - \gamma_0)^\omega \rangle$ be of Type 4. When $\omega = 0$, we have $u \in \mathcal{C}_4$, which implies that $d_H(\mathcal{C}_4) = 1$.

Now let $\omega \geq 1$. And let $c(x)$ be an arbitrary nonzero element of $\mathcal{C}_4$, then $c(x)$ can be represented as

$$c(x) = (f(x) + ug(x))((x - \gamma_0)^\delta + u(x - \gamma_0)^t h(x))$$
$$+ (a(x) + ub(x))(u(x - \gamma_0)^\omega),$$

where $f(x), g(x), a(x), b(x) \in \mathbb{F}_{p^m}[x]$. We consider two cases.

∗ **Case 1:** $f(x) = 0$, then we can written $c(x)$ as

$$c(x) = u(x - \gamma_0)^\omega (g(x)(x - \gamma_0)^{\delta-\omega} + a(x))$$

and therefore we have

$$wt_H(c(x)) \geq d_H(\langle u(x - \gamma_0)^\omega \rangle)$$

∗ **Case 2:** $f(x) \neq 0$. Then we have

$$c(x) = f(x)(x - \gamma_0)^\delta + u[g(x)(x - \gamma_0)^\delta + f(x)(x - \gamma_0)^t h(x) \\ + a(x)(x - \gamma_0)^\omega],$$

which, by (1), implies that

$$wt_H(c(x)) \geq wt_H(f(x)(x - \gamma_0)^\delta) \\ \geq d_H(\langle(x - \gamma_0)^\delta\rangle_{\mathbb{F}})$$

Because, $\langle(x - \gamma_0)^\delta\rangle \subseteq \langle(x - \gamma_0)^\omega\rangle$, we have

$$d_H(\langle(x - \gamma_0)^\delta\rangle_{\mathbb{F}}) \geq d_H(\langle(x - \gamma_0)^\omega\rangle_{\mathbb{F}}).$$

From this, we get $wt_H(c(x)) \geq d_H(\langle(x - \gamma_0)^\omega\rangle_{\mathbb{F}})$ for each $c(x)$ nonzero element of $\mathcal{C}_4$. This implies that

$$d_H(\mathcal{C}_4) \geq d_H(\langle(x - \gamma_0)^\omega\rangle_{\mathbb{F}}) \tag{4}$$

Further, as $\langle u(x - \gamma_0)^\omega\rangle \subseteq \mathcal{C}_4$, then

$$d_H(\langle(x - \gamma_0)^\omega\rangle_{\mathbb{F}}) = d_H(\langle u(x - \gamma_0)^\omega\rangle) \geq d_H(\mathcal{C}_4) \tag{5}$$

By (4) and (5), we get

$$d_H(\mathcal{C}_4) = d_H(\langle(x - \gamma_0)^\omega\rangle_{\mathbb{F}}).$$

Finally, by the above assertion and by applying Theorem 1, we get the desired result. □

Remark 1. Theorems 4 and 5 are a correct version of a Theorems 3.3, 3.4 in [7] and Theorem 5.2 (Type 3 and 4) in [12] which are not completely true, so we had to add some assumptions to correct them completely.

5   Symbol-Pair Distance

Our technique in Sect. 4 can also be extended to compute the symbol-pair distance of the ideals of the ring $\mathcal{R}_\gamma$ (Type 3 and 4). Type 1 consists of the trivial ideals. It is easy to see that $d_{sp}(\mathcal{C}) = 0$ when $\mathcal{C} = \{0\}$, while $d_{sp}(\mathcal{C}) = 2$ when $\mathcal{C} = \{1\}$. The symbol-pair distances of Type 2 were provided in [7].

We first observe that

$$wt_{sp}(a(x) + ub(x)) \geq wt_{sp}(a(x)), \tag{6}$$

where $a(x), b(x) \in \mathbb{F}_{p^m}[x]/\langle x^{p^s} - \gamma\rangle$.

In [7, Theorem 4.3(Type3)], Dinh et al. stated that: $d_{sp}(\mathcal{C}_3) = d_{sp}(\langle(x - \gamma_0)^\delta\rangle_{\mathbb{F}})$. This result is not always true, which we illustrate in the following example.

Example 2. Let $p = 3$, $s = 2$ and $\gamma = 1$. Consider the cyclic code $\mathcal{C}_3 = \langle (x - 1)^7 + u(x - 1)^3 h(x) \rangle$ of length 3^2 over $\mathcal{R} = \mathbb{F}_{3^m} + u\mathbb{F}_{3^m}$, where $h(x) \neq 0$. Here $\delta = 7$ and $t = 3$. Then $\mathsf{T} = 5$, which implies that $\langle u(x - 1)^5 \rangle \subseteq \mathcal{C}_3$. This implies that $d_{sp}(\langle u(x - 1)^5 \rangle) = d_{sp}(\langle (x - 1)^5 \rangle_{\mathbb{F}}) \geq d_{sp}(\mathcal{C}_3)$. By Theorem 2, we see that $d_{sp}(x - 1)^5 \rangle_{\mathbb{F}}) = 6$ and $d_{sp}(x - 1)^7 \rangle_{\mathbb{F}}) = 9$. Now, we see that $d_{sp}(\mathcal{C}_3) \neq d_{sp}(\langle (x - 1)^7 \rangle_{\mathbb{F}})$. This example shows that [7, Theorem 4.3 (Type3)] is incorrect in general.

In the following theorem, we correct [7, Theorem 4.3 (Type3)] as follows.

Theorem 6. *Let $\mathcal{C}_3 = \langle (x - \gamma_0)^\delta + u(x - \gamma_0)^t h(x) \rangle$ be a γ-constacyclic codes of length p^s over $\mathcal{R}$ of* Type 3 *(as classified in Theorem 3), where $1 \leq \mathsf{T} \leq \delta \leq p^s - 1$, $0 \leq t < \mathsf{T}$, either $h(x)$ is 0 or $h(x)$ is a unit and T is the smallest integer satisfying $u(x - \gamma_0)^{\mathsf{T}} \in \mathcal{C}_3$, i.e.,*

$$\mathsf{T} = \begin{cases} \delta, & if\, h(x) = 0, \\ \min\{\delta, p^s - \delta + t\}, & if\, h(x) \neq 0. \end{cases}$$

Then the symbol-pair distance $d_{sp}(\mathcal{C}_3)$ of the code $\mathcal{C}_3$ is given by

$$d_{sp}(\mathcal{C}_3) = \begin{cases} \bullet\ 3p^k, & if\ \mathsf{T} = p^s - p^{s-k} + 1, \\ \quad where\ 0 \leq k \leq s - 2, \\ \\ \bullet\ 4p^k, & if \\ p^s - p^{s-k} + 2 \leq \mathsf{T} \leq p^s - p^{s-k} + p^{s-k-1}, \\ \quad where\ 0 \leq k \leq s - 2, \\ \\ \bullet\ 2(\sigma + 2)p^k, & if \\ p^s - pr + \sigma r + 1 \leq \mathsf{T} \leq p^s - pr + (\sigma + 1)r, \\ \quad where\ r = p^{s-k-1},\ 0 \leq k \leq s - 2 \\ \quad and\ 1 \leq \sigma \leq p - 2, \\ \\ \bullet\ (\sigma + 2)p^{s-1}, & if\ \mathsf{T} = p^s - p + \sigma, \\ \quad where\ 1 \leq \sigma \leq p - 2, \\ \\ \bullet\ p^s, & if\ \mathsf{T} = p^s - 1. \end{cases}$$

Proof. Let $\mathcal{C}_3 = \langle (x - \gamma_0)^\delta + u(x - \gamma_0)^t h(x) \rangle$ be of Type 3. And let $c(x)$ be an arbitrary nonzero element of $\mathcal{C}_3$, then $c(x)$ can be written as

$$c(x) = [f(x) + ug(x)][(x - \gamma_0)^\delta + u(x - \gamma_0)^t h(x)],$$

where $f(x), g(x) \in \mathbb{F}_{p^m}[x]$. We consider two cases.

　∗ Case 1: $f(x) = 0$. In this case, $g(x) \neq 0$. We have $c(x) = ug(x)(x - \gamma_0)^\delta$, which implies that

$$\begin{aligned} wt_{sp}(c(x)) &= wt_{sp}(ug(x)(x - \gamma_0)^\delta) \\ &\geq d_{sp}(\langle u(x - \gamma_0)^\delta \rangle) = d_{sp}(\langle (x - \gamma_0)^\delta \rangle_{\mathbb{F}}). \end{aligned}$$

Since, $\langle (x - \gamma_0)^\delta \rangle \subseteq \langle (x - \gamma_0)^\mathsf{T} \rangle$, we have

$$d_{sp}(\langle (x - \gamma_0)^\delta \rangle_\mathbb{F}) \geq d_{sp}(\langle (x - \gamma_0)^\mathsf{T} \rangle_\mathbb{F}).$$

$*$ Case 2: $f(x) \neq 0$. Then we have

$$c(x) = f(x)(x - \gamma_0)^\delta + u[g(x)(x - \gamma_0)^\delta + f(x)(x - \gamma_0)^t h(x)].$$

By (6), we obtain that

$$\begin{aligned} wt_{sp}(c(x)) &\geq wt_{sp}(f(x)(x - \gamma_0)^\delta) \\ &\geq d_{sp}(\langle (x - \gamma_0)^\delta \rangle_\mathbb{F}) \\ &\geq d_{sp}(\langle (x - \gamma_0)^\mathsf{T} \rangle_\mathbb{F}). \end{aligned}$$

From this, we get $wt_{sp}(c(x)) \geq d_{sp}(\langle (x - \gamma_0)^\mathsf{T} \rangle_\mathbb{F})$ for each $c(x)$ nonzero element of $\mathcal{C}_3$. This implies that

$$d_{sp}(\mathcal{C}_3) \geq d_{sp}(\langle (x - \gamma_0)^\mathsf{T} \rangle_\mathbb{F})$$

On the other hand we have that

$$\langle u(x - \gamma_0) \rangle \subseteq \mathcal{C}_3$$

then $d_{sp}(\langle (x - \gamma_0)^\mathsf{T} \rangle_\mathbb{F}) = d_{sp}(\langle u(x - \gamma_0)^\mathsf{T} \rangle) \geq d_{sp}(\mathcal{C}_3)$ and we get $d_{sp}(\mathcal{C}_3) = d_{sp}(\langle (x - \gamma_0)^\mathsf{T} \rangle_\mathbb{F})$. The rest of the proof follows from Theorem 2 and the discussion above. $\qquad\square$

Corollary 2. *Following the same notations as in Theorem 6, we have the following:*

1. If $h(x)$ is 0 or $h(x)$ is a unit and $1 \leq \delta \leq \frac{p^s + t}{2}$, then

$$d_{sp}(\mathcal{C}_3) = \begin{cases} \bullet\ 3p^k, \ \ if\ \delta = p^s - p^{s-k} + 1, \\ \quad where\ 0 \leq k \leq s - 2, \\[2mm] \bullet\ 4p^k, \ \ if \\ p^s - p^{s-k} + 2 \leq \delta \leq p^s - p^{s-k} + p^{s-k-1}, \\ \quad where\ 0 \leq k \leq s - 2, \\[2mm] \bullet\ 2(\sigma + 2)p^k, \ \ if \\ p^s - pr + \sigma r + 1 \leq \delta \leq p^s - pr + (\sigma + 1)r, \\ \quad where\ \ r = p^{s-k-1}, \ \ 0 \leq k \leq s - 2 \\ \quad and\ \ 1 \leq \sigma \leq p - 2, \\[2mm] \bullet\ (\sigma + 2)p^{s-1}, \ \ if\ \delta = p^s - p + \sigma, \\ \quad where\ \ 1 \leq \sigma \leq p - 2, \\[2mm] \bullet\ p^s, \ if\ \delta = p^s - 1. \end{cases}$$

2. If $h(x)$ is a unit and $\frac{p^s+t}{2} < \delta \leq p^s - 1$, then

$$d_{sp}(\mathcal{C}_3) = \begin{cases} \bullet\ 3p^k, \ \ if\ \delta = t + p^{s-k} - 1, \\ \quad\ where\ 0 \leq k \leq s - 2, \\ \\ \bullet\ 4p^k, \ if \\ \quad t + p^{s-k} - p^{s-k-1} \leq \delta \leq t + p^{s-k} - 2, \\ \quad\ where\ 0 \leq k \leq s - 2, \\ \\ \bullet\ 2(\sigma + 2)p^k, \ \ if \\ \quad t + pr - (\sigma + 1)r \leq \delta \leq t + pr - \sigma r - 1, \\ \quad\ where\ \ r = p^{s-k-1}, \ \ 0 \leq k \leq s - 2 \\ \quad\quad\ and\ \ 1 \leq \sigma \leq p - 2, \\ \\ \bullet\ (\sigma + 2)p^{s-1}, \ \ if\ \delta = t + p - \sigma, \\ \quad\ where\ \ 1 \leq \sigma \leq p - 2, \\ \\ \bullet\ p^s, \ if\ \ \delta = t + 1. \end{cases}$$

Theorem 7. *Let* $\mathcal{C}_4 = \langle (x - \gamma_0)^\delta + u(x - \gamma_0)^t h(x), u(x - \gamma_0)^\omega \rangle$ *be a* γ-*constacyclic codes of length* p^s *over* $\mathcal{R}$ *of* Type 4 *(as determined in Theorem 3), where* $0 \leq \omega < \delta \leq p^s - 1$ *and either* $h(x)$ *is* 0 *or* $h(x)$ *is a unit. Then*

$$d_{sp}(\mathcal{C}_4) = \begin{cases} \bullet\ 2, \ \ if\ \omega = 0, \\ \\ \bullet\ 3p^k, \ \ if\ \omega = p^s - p^{s-k} + 1, \\ \quad\ where\ 0 \leq k \leq s - 2, \\ \\ \bullet\ 4p^k, \ \ if \\ \quad p^s - p^{s-k} + 2 \leq \omega \leq p^s - p^{s-k} + p^{s-k-1}, \\ \quad\ where\ 0 \leq k \leq s - 2, \\ \\ \bullet\ 2(\sigma + 2)p^k, \ \ if \\ \quad p^s - pr + \sigma r + 1 \leq \omega \leq p^s - pr + (\sigma + 1)r, \\ \quad\ where\ \ r = p^{s-k-1}, \ \ 0 \leq k \leq s - 2 \\ \quad\quad\ and\ \ 1 \leq \sigma \leq p - 2, \\ \\ \bullet\ (\sigma + 2)p^{s-1}, \ \ if\ \ \omega = p^s - p + \sigma, \\ \quad\ where\ \ 1 \leq \sigma \leq p - 2. \end{cases}$$

Proof. Let $\mathcal{C}_4 = \langle (x - \gamma_0)^\delta + u(x - \gamma_0)^t h(x), u(x - \gamma_0)^\omega \rangle$ be of Type 4. When $\omega = 0$, we have $u \in \mathcal{C}_4$, which implies that $d_{sp}(\mathcal{C}_4) = 2$.

Now let $\omega \geq 1$. And let $c(x)$ be an arbitrary nonzero element of $\mathcal{C}_4$, then $c(x)$ can be represented as

$$c(x) = (f(x) + ug(x))((x - \gamma_0)^\delta + u(x - \gamma_0)^t h(x)) \\ + (a(x) + ub(x))(u(x - \gamma_0)^\omega),$$

where $f(x), g(x), a(x), b(x) \in \mathbb{F}_{p^m}[x]$. We consider two cases.

∗ Case 1: $f(x) = 0$, then we can written $c(x)$ as

$$c(x) = u(x - \gamma_0)^\omega (g(x)(x - \gamma_0)^{\delta - \omega} + a(x)),$$

and therefore we have

$$wt_{sp}(c(x)) \geq d_{sp}(\langle u(x - \gamma_0)^\omega \rangle).$$

∗ Case 2: $f(x) \neq 0$. Then we have

$$\begin{aligned} c(x) = {} & f(x)(x - \gamma_0)^\delta + u[g(x)(x - \gamma_0)^\delta + f(x)(x - \gamma_0)^t h(x) \\ & + a(x)(x - \gamma_0)^\omega], \end{aligned}$$

which, by (6), implies that

$$\begin{aligned} wt_{sp}(c(x)) & \geq wt_{sp}(f(x)(x - \gamma_0)^\delta) \\ & \geq d_{sp}(\langle (x - \gamma_0)^\delta \rangle_\mathbb{F}). \end{aligned}$$

Because, $\langle (x - \gamma_0)^\delta \rangle \subseteq \langle (x - \gamma_0)^\omega \rangle$, we have

$$d_{sp}(\langle (x - \gamma_0)^\delta \rangle_\mathbb{F}) \geq d_{sp}(\langle (x - \gamma_0)^\omega \rangle_\mathbb{F}).$$

From this, we get $wt_{sp}(c(x)) \geq d_{sp}(\langle (x - \gamma_0)^\omega \rangle_\mathbb{F})$ for each $c(x)$ nonzero element of $\mathcal{C}_4$. This implies that

$$d_{sp}(\mathcal{C}_4) \geq d_{sp}(\langle (x - \gamma_0)^\omega \rangle_\mathbb{F}).$$

Further, as $\langle u(x - \gamma_0)^\omega \rangle \subseteq \mathcal{C}_4$, then $d_{sp}(\langle (x - \gamma_0)^\omega \rangle_\mathbb{F}) = d_{sp}(\langle u(x - \gamma_0)^\omega \rangle) \geq d_{sp}(\mathcal{C}_4)$ and we get $d_{sp}(\mathcal{C}_4) = d_{sp}(\langle (x - \gamma_0)^\omega \rangle_\mathbb{F})$. The rest of the proof follows from Theorem 2 and the discussion above. □

Remark 2. Theorems 6 and 7 are a correct version of a Theorem 4.3 (Type 3 and 4) in [7] which is not entirely true, so we had to add some hypotheses to correct it completely.

Example 3. γ-constacyclic codes of length 9 over the chain ring $\mathcal{R} = \mathbb{F}_3 + u\mathbb{F}_3$ are precisely the ideals of $\mathcal{R}[x]/\langle x^9 - \gamma \rangle$, where $\gamma \in \{1, 2\}$. The following Table 1 shows the representation of all γ-constacyclic codes of length 9 over the chain ring $\mathbb{F}_3 + u\mathbb{F}_3$ of Type 3, where $h(x)$ is a unit, $0 \leq t < \delta$ and $\frac{9 + t}{2} < \delta \leq 8$, together with their Hamming distances d$_\mathsf{H}$ and their symbol-pair distances d$_\mathsf{sp}$.

Example 4. Cyclic codes of length 3 over the chain ring $\mathcal{R} = \mathbb{Z}_3 + u\mathbb{Z}_3$ are precisely the ideals of $\mathcal{R}[x]/\langle x^3 - 1 \rangle$. The following Table 2 gives all cyclic codes of length 3 over the chain ring $\mathbb{Z}_3 + u\mathbb{Z}_3$ of Type 4, together with their Hamming distances d$_\mathsf{H}$ and their symbol-pair distances d$_\mathsf{sp}$.

Table 1. γ-constacyclic codes of length **9** over the chain ring $\mathbb{F}_3 + u\mathbb{F}_3$ of Type 3 ($\mathbf{h(x)}$ is a unit, $0 \le t < \delta$ and $\frac{9+t}{2} < \delta \le 8$).

Ideal	d_H	d_{sp}
$\rightarrow \delta = 5$ and $t = 0$:		
$*\langle (x-\gamma)^5 + h_0 u + h_1 u(x-\gamma) + h_2 u(x-\gamma)^2 + h_3 u(x-\gamma)^3 \rangle$	3	6
$\rightarrow \delta = 6$ and $t = 0$:		
$*\langle (x-\gamma)^6 + h_0 u + h_1 u(x-\gamma) + h_2 u(x-\gamma)^2 \rangle$	2	4
$\rightarrow \delta = 6$ and $t = 1$:		
$*\langle (x-\gamma)^6 + h_0 u(x-\gamma) + h_1 u(x-\gamma)^2 + h_2 u(x-\gamma)^3 \rangle$	3	6
$\rightarrow \delta = 6$ and $t = 2$:		
$*\langle (x-\gamma)^6 + h_0 u(x-\gamma)^2 + h_1 u(x-\gamma)^3 + h_2 u(x-\gamma)^4 \rangle$	3	6
$\rightarrow \delta = 7$ and $t = 0$:		
$*\langle (x-\gamma)^7 + h_0 u + h_1 u(x-\gamma) \rangle$	2	4
$\rightarrow \delta = 7$ and $t = 1$:		
$*\langle (x-\gamma)^7 + h_0 u(x-\gamma) + h_1 u(x-\gamma)^2 \rangle$	2	4
$\rightarrow \delta = 7$ and $t = 2$:		
$*\langle (x-\gamma)^7 + h_0 u(x-\gamma)^2 + h_1 u(x-\gamma)^3 \rangle$	3	6
$\rightarrow \delta = 7$ and $t = 3$:		
$*\langle (x-\gamma)^7 + h_0 u(x-\gamma)^3 + h_1 u(x-\gamma)^4 \rangle$	3	6
$\rightarrow \delta = 7$ and $t = 4$:		
$*\langle (x-\gamma)^7 + h_0 u(x-\gamma)^4 + h_1 u(x-\gamma)^5 \rangle$	3	6
$\rightarrow \delta = 8$ and $t = 0$:		
$*\langle (x-\gamma)^8 + h_0 u \rangle$	2	3
$\rightarrow \delta = 8$ and $t = 1$:		
$*\langle (x-\gamma)^8 + h_0 u(x-\gamma) \rangle$	2	4
$\rightarrow \delta = 8$ and $t = 2$:		
$*\langle (x-\gamma)^8 + h_0 u(x-\gamma)^2 \rangle$	2	4
$\rightarrow \delta = 8$ and $t = 3$:		
$*\langle (x-\gamma)^8 + h_0 u(x-\gamma)^3 \rangle$	3	6
$\rightarrow \delta = 8$ and $t = 4$:		
$*\langle (x-\gamma)^8 + h_0 u(x-\gamma)^4 \rangle$	3	6
$\rightarrow \delta = 8$ and $t = 5$:		
$*\langle (x-\gamma)^8 + h_0 u(x-\gamma)^5 \rangle$	3	6
$\rightarrow \delta = 8$ and $t = 6$:		
$*\langle (x-\gamma)^8 + h_0 u(x-\gamma)^6 \rangle$	6	9
where $h_0 \in \{1,2\}, h_1, h_2, h_3 \in \{0,1,2\}$.		

Table 2. Cyclic codes of length **3** over $\mathbb{Z}_3 + u\mathbb{Z}_3$ of Type 4.

Ideal	d_H	d_{sp}
$*\langle(x-1), u\rangle$	1	2
$*\langle(x-1)^2, u\rangle$	1	2
$*\langle(x-1)^2, u(x-1)\rangle$	2	3

6   Conclusion

Let p be a prime, s, m be positive integers and γ be a nonzero element of $\mathbb{F}_{p^m}$. Let $\mathcal{R} = \mathbb{F}_{p^m}[u]/\langle u^2\rangle$ be the finite commutative chain ring with unity. In this paper, The structure of γ-constacyclic codes of length p^s over $\mathcal{R}$ is studied. We also completed the problem of determination of the Hamming and symbol-pair distance of γ-constacyclic codes of length p^s over $\mathcal{R}$.

Acknowledgment. The authors would like to thank the reviewers and the editor for their valuable comments and suggestions, which have greatly improved the quality of this paper.

References

1. Cassuto, Y., Blaum, M.: Codes for symbol-pair read channels. In: Proceedings of International Symposium Information Theory, Austin, TX, USA, pp. 988–992 (2010)
2. Cassuto, Y., Blaum, M.: Codes for symbol-pair read channels. IEEE Trans. Inf. Theory **57**(12), 8011–8020 (2011)
3. Cassuto, Y., Litsyn, S.: Symbol-pair codes: algebraic constructions and asymptotic bounds. In: Proceedings of International Symposium Information Theory, St. Petersburg, Russia, pp. 2348–2352 (2011)
4. Dinh, H.Q.: On the linear ordering of some classes of negacyclic and cyclic codes and their distance distributions. Finite Fields Appl. **14**, 22–40 (2008)
5. Dinh, H.Q.: Constacyclic codes of length p^s over $\mathbb{F}_{p^m} + u\mathbb{F}_{p^m}$. J. Algebra **324**, 940–950 (2010)
6. Dinh, H.Q., López-Permouth, S.R.: Cyclic and negacyclic codes over finite chain rings. IEEE Trans. Inform. Theory **50**, 1728–1744 (2004)
7. Dinh, H.Q., Nguyen, B.T., Singh, A.K., Sriboonchitta, S.: Hamming and symbol-pair distances of repeated-root constacyclic codes of prime power lengths over $\mathbb{F}_{p^m} + u\mathbb{F}_{p^m}$. IEEE Commun. Lett. **22**, 2400–2403 (2018)
8. Dinh, H.Q., Nguyen, B.T., Singh, A.K., Sriboonchitta, S.: On the symbol-pair distance of repeated-root constacyclic codes of prime power lengths. IEEE Trans. Inform. Theory **64**, 2417–2430 (2018)
9. Hirotomo, M., Takita, M., Morii, M.: Syndrome decoding of symbol-pair codes. In: Proceedings of Information Theory Workshop, Hobart, TAS, Australia, pp. 162–166 (2014)
10. Huffman, W.C., Pless, V.: Fundamentals of Error-Correcting Codes. Cambridge University Press, Cambridge (2003)

11. Kai, X., Zhu, S., Li, P.: A construction of new MDS symbol-pair codes. IEEE Trans. Inf. Theory **61**(11), 5828–5834 (2015)
12. Liu H., Youcef, M.: A note on hamming distance of constacyclic codes of length p^s over $\mathbb{F}_{p^m} + u\mathbb{F}_{p^m}$. arXiv:1612.03731v1 (2016)
13. MacWilliams, F.J., Sloane, N.J.A.: The Theory of Error-Correcting Codes. 10th Impression, North-Holland, Amsterdam (1998)
14. Yaakobi, E., Bruck, J., Siegel, P.H.: Decoding of cyclic codes over symbol-pair read channels. In: Proceedings of International Symposium Information Theory, Cambridge, MA, USA, pp. 2891–2895 (2012)

Intrusion Detection Techniques

A Comparison of Different Machine Learning Algorithms for Intrusion Detection

Basma Karbal$^{(\boxtimes)}$ and Rahal Romadi

Laboratory of Research, IRDA Team, ENSIAS,
Mohammed V University, Rabat, Morocco
{basma.karbal,r.romadi}@um5s.net.ma

Abstract. With the rapid development of the internet, intrusion detection became one of the major research problems in computer security. Many Intrusion Detection Systems (IDS) use data mining algorithms for classifying network traffic data and detecting different security violations. In this paper, we present some of the datasets and methods employed with the focus on network anomaly detection. We compare different machine learning techniques used in the latest research carried out for developing network intrusion detection systems. We also present an overview of some deep learning methodologies and their application for IDS purposes. The primary objective of this survey is to provide with a researcher, the state of the artwork already performed in this field of research.

Keywords: Intrusion detection · Data mining · Deep learning · Network anomaly detection

1 Introduction

For every organization, information security issues have always been important to protect information from unauthorized access. The concept of intrusion can be expressed as the set of actions that attempts to invade a system. They attempt to compromise a major part of the system such as confidentiality, availability or integrity. They try to have access to the information, to manipulate the data so that the information becomes unusable or unreliable in the system.

Intrusion detection is defined as an ensemble of hardware and software components that detect malicious behavior. Intrusion detection system scan activities and hence can find indications of intrusions. Traditionally, there are two categories of IDS techniques: Anomaly detection and misuse detection. Anomaly detection is based on deviations from normal conduct based on a system. Unknown intrusion can be detected while misuse or signature detection recognizes intrusions based on known patterns of malicious activity, called attack signatures. But misuse detection can only detect attacks that contain a known pattern, unlike the anomaly-based system. Although a signature or misuse based

© Springer Nature Switzerland AG 2020
M. Belkasmi et al. (Eds.): ACOSIS 2019, CCIS 1264, pp. 157–169, 2020.
https://doi.org/10.1007/978-3-030-61143-9_13

design has the advantage of identifying attacks with high accuracy and tending to produce less false alarms than the anomaly-based.

Therefore, recent research is founded on modeling the comportment and use data mining algorithms, artificial intelligence techniques, and statistical analysis to identify attacks. This article cites some data preprocessing to manage large datasets with a high dimension for developing IDS. The rest of the paper is formed as follows. Section 2 exposes related work based on intrusion detection research, Sect. 3 describes the intrusion detection datasets, Sect. 4 explores different data preprocessing techniques, Sect. 5 presents a detailed analysis of NSL-KDD on numerous data mining algorithms. Section 6 explains the experimental results obtained by the researchers. Section 7 describes the effectiveness of deep learning for network anomaly detection. The conclusion and future work are discussed in the last section.

2   Related Work

Different researchers applied various data mining techniques for classification data to develop an intrusion detection system (IDS).

Amudha et al. [6] performed a study using data mining classification algorithm namely Naïve Bayes, Random Forest, J48 and NB Tree in KDD Cup 99 dataset.

Panda et al. [22] compared many algorithms for classification namely ID3, Naïve Bayes and J48 for intrusion detection. They were evaluated using 10-fold cross-validation test.

Kumar et al. [19] presented a study on anomaly and misuse attack detection using a decision tree algorithm.

Kohavi et al. [15] proposed a new hybrid algorithm called Decision Tree Hybrid for scaling up the accuracy of Naïve Bayes.

Zhang et al. [31] have introduced Hidden Naïve Bayes models. They create hidden parents using weighted one dependence estimators and they added hidden parents for each attribute on Naïve Bayes.

Reza et al. [23] investigated the effect of Synthetic Minority Oversampling Technique (SMOTE) combined with the Cluster Center And Nearest Neighbor (CANN). They proposed a hybrid approach to increase the detection rate of infrequent attacks such as U2R (User to Root) and R2L (Remove to Local).

[3] used Self-Taught learning (STL) as a proposed deep learning approach for Intrusion Detection Systems. This effective approach is formed by the combination of the Support Vector Machine (SVM) and the Sparse Auto-Encoder (SAE). Testing and training time is significantly reduced, and the accuracy of SVM predictions is effectively improved. The dataset NSL-KDD was used for experimentation and to compare the deep learning framework of the STL with the single support vector machine and with other classification methods such as Random Forest, Naïve Bayesian, Multilayer Perceptron etc. for binary and multi-class classification. The proposed framework consists of two steps. In the first step, unsupervised feature learning is performed without using labels.

In the next step, the representation is combined with labelled data, and the Support Vector Machine is used for classification. Experimental results show that the proposed approach produced an accuracy of 99.423% for binary classification and 99.414% for multi-class classification. This paper [32] examines the possibility of using a deep learning model for network intrusion based on Denoising Auto-Encoder (DAE). The dimensionality of the features is reduced by using a weight loss function to select a limited number of important features. They then used the Multi-Layer Perceptron (MLP) for classification. For the data set, UNSW-NB was used. The proposed approach resulted in an improved accuracy of 98.80% with an F-score of 0.952. 10 features were selected out of 202 features with a feature selection ratio of 5.9%.

3 Intrusion Detection Dataset

The KDD99 has been used as a reference for intrusion detection to evaluate the performance of anomaly detection techniques. It was collected by MIT, Lincoln Lab by the DARPA Intrusion Detection Evaluation Program in 1998. It is approximately nine weeks of raw TCP dump data captured for a simulated U.S. Air Force LAN [7]. Various statistical analysis [28] reveal the inherent drawbacks of the KDD99 Cup. Tavallaee et al. observed that 78% and 75% of redundant records are included in the training and test data set. NSL-KDD [1] is a refined version of its predecessor. It is proposed to fix some of the inheritance problems:

- Duplicate records are deleted in the training set and the test set to prevent the classifier from generating an error-free result during performance.
- For each difficulty level group, the selection of records is inversely proportional to the number of records in the original KDD dataset.
- The number of records available in the training and test datasets makes it possible to run algorithms on the entire dataset.

The Table 1 reveals the main attacks in the train and test datasets. All attacks fall into four categories:

- **U2R** *(User to Root)*: an attacker first accesses a local machine and tries to access superuser privileges.
- **DoS** *(Denial of Service)*: an attacker makes memory resources or certain computers too full to prevent legitimate users from accessing them.
- **R2L** *(Remote to Local)*: an attacker attempts to access a computer or device from the outside by compromising security.
- **Probing**: an attacker collects details about the victim machine over the network.

Each sample of the NSL-KDD dataset represents a connection between two network hosts. It consists of 41 attributes, including 32 numerical attributes, 6 binary attributes and 3 categorical. A label is assigned to each record as a normal or specific attack.

Table 1. Type of attacks grouped into four categories

Attack class	Attack type
DoS	Back/Land/Neptune/Pod/Smurf/Teardrop/Appache2/Dosnuke /Tcp-reset/Mailbomb/Selfping/Processtable/Udpstorm/Warezclient/Worm
Probe	Portsweep/Ipsweep/Queso/Satan/Mscan/Nmap/Saint/Ntinfoscan /Isdomain/Illegal-sniffer
R2L	Guess_password/Ftp_write/Imap/Phf/Multihop/Warezmaster/Dict /Netcat/Sendmail/Ncftp/Xlock/Xsnoop/Framespof/Ppmacro/Guest /Netbus/Snmpget/Httptunnel/Named
U2R	Buffer_overflow/Loadmodule/RootKit/Perl/Sqlattack/Xterm/Ps /Sechole/Eject/Nukepw/Secret/Yaga/Fdformat/Ffbconfig/Casesen /Ntfsdos/Ppmacro

Consequently, NSL-KDD dataset is better than the original KDD99 dataset, which allows the evaluation results of different learning classifiers to be consistent and comparable. In addition, some arguments against the use of this data set are listed by McHugh [20]. The proposed dataset still suffers from some problems and cannot be an ideal representative of existing real networks, due to the insufficiency of the public dataset for networked IDS. However, it can still be used as an effective basic data set for the evaluation of the Intrusion Detection System.

4   Methods and Materials

This section presents an overview of the methods and materials used in the construction of the Intrusion Detection System (IDS) by the researchers. All the tools used to improve the performance of the different IDS and to form the model.

4.1   Data Preprocessing

The data from network traffic is very large and when we process the data directly it takes time and the evaluation may not be accurate, so we should opt for pre-processing the data. The main purpose is to transform the original input data into a form suitable for analysis. Data pre-processing includes character value to numeric value conversion, normalization, discretization and dimensionality reduction. Figure 1 illustrates the block diagram of data preprocessing and data flow.

Conversion Characters to Numerical Values. Some classifiers require input data to be presented in numerical form to improve accuracy. For our database, there are three features with character values: Service/Flag/Protocol Type. To convert them into numerical values, we calculate the occurrence of each feature and then order these values in ascending order. For the attribute with the highest number of repetitions, take the value 1 and for the least frequent ones, take the value 2... etc.

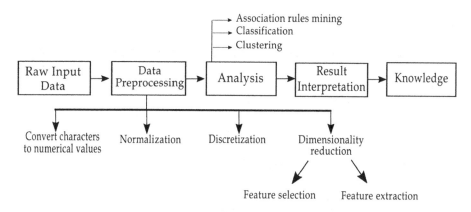

Fig. 1. Data flow

Normalization. Many features contain a range of different values. Therefore, normalizing or scaling is an important step before selecting features to give accurate results. Normalization is a processing technique used to resize attribute values to fit within a specific range. Many scaling methods are available, but the most commonly used are:

- **Min-max normalization:** The attributes are scaled from 0 to 1 using the Eq. (1) . With x_i is the value of the element, $min(x)$ is the minimum and $max(x)$ is the maximum value. It ensures that all features will have exactly the same scale, but does not handle outliers well.

$$newX_i = \frac{X_i - min(x)}{max(x) - min(x)} \tag{1}$$

- **Z-score normalization:** This technique takes the advantage over min-max normalization when the attribute has a negative variance. The z-score transforms the data such that the standard deviation of the new value is equal to 1 and the mean is equal to 0 using (2). Where X_i is the current sample, X is the mean of all data points in the sample, and σ is the standard deviation. It handles outliers, but does not produce normalized data with exactly the same scale.

$$newX_i = \frac{X_i - \bar{X}_s}{\sigma_{X,s}} \tag{2}$$

Discretization. Discretization algorithms [18] is one of the techniques used for pre-processing network data. It involves transforming the continuous features of a data set into discrete features. It is mainly used to optimize intrusion detection performance and to increase the classification capability between abnormal and normal behavior. There are two common discretization mechanisms [2,11]:

- **Supervised methods:** use the class information
- **Unsupervised methods:** don't use any class information

Supervised discretization involves the arrangement of continuous attributes in several disjoint (discrete) intervals using the target variable (i.e. the class attribute). It is based on finding the correct intervals caused by the cut-off points.

There exist many supervised discretization techniques in the literature including the statistics-based methods [10], entropy-based method [14] and the error-based method [17]. But the most common supervised discretization technique is the Entropy-Based Discretization (EBD) proposed by Fayyad and Irani [14]. This technique uses entropy measurement as a standard criterion to recursively partition the values of a continuous candidate and the Minimum Description Length (MDL) criterion to stop discretization.

Unsupervised discretization methods rely on dividing the continuous values into K intervals, where K is a parameter specified by the user. There are many representative algorithms of unsupervised discretization:

- **Equal Width Discretization (EWD):** EWD divides the values between the minimum and maximum values of the discretized attribute into K intervals of equal width (Where K is given by the user).
- **Equal Frequency Discretization (EFD):** EFD determines the minimum and the maximum values of the discretized attribute and divides the sorted values into K intervals (K is determined by the user).
- **Proportional K-Interval Discretization (PKID):** is introduced by Yang et al. [29] in order to improve the performance of Naïve Bayes classifier. The discretization bias in the PKID is related to the size of the interval and the variance of discretization with respect to the interval number. Therefore, the smaller the interval size indicates the lower the bias but the higher the variance. Thus, by adjusting the number and size of intervals, an appropriate trade-off between the bias and variance can be found and hence it can achieve lower learning error.

Dimensionality Reduction. Two approaches are available for dimensionality reduction: Feature selection and Feature extraction. But before applying dimensionality reduction, one attribute namely *num_outbound_cmds* should be eliminated from the data because his value is equal to zero for all records.

- **Feature selection** is important for machine learning algorithm performance. It concerns the diminution of features in order to represent the whole dataset and progress the performance of the data analysis with achieving a higher accuracy rate [12]. There are two main categories for selecting features: filtering and wrapping methods [16].
  The selection methods for filter features consider the feature independently or in relation to the dependent variable. The methods use a statistical measure to score each feature. After ranking each score, they select the feature with independence from any predictors. Commonly known metrics for filtering methods are the chi-square test [27], information gain [25] and correlation coefficient scores [9].
  Feature selection with wrapper method creates all possible subsets of features from the feature vector. Then, a predictive model induced from the features

in each subset and used to evaluate the combination of the features and the one that leads to the optimum performance of the predictor is selected.

- **Feature extraction** transforms existing features into a smaller space that still captures most of the useful information. Some algorithms have already incorporated feature extraction, such as deep learning, which generates an increasingly valuable picture of the raw data entered by each hidden neural layer. Feature extraction can be either unsupervised (i.e. Principal Component Analysis) or supervised (i.e. Linear Discriminant Analysis). PCA is an unsupervised algorithm that takes our original data as input and tries to find a combination of the input features that can best summarize the distribution of the original data to reduce its dimensions. PCA is quickly and easily implemented. But it is a manually set or tuned threshold for an explained cumulative variance. Unlike PCA, which maximizes the explained variance, LDA maximizes separability between classes and is a supervised method that can only be used with labeled data. LDA requires labeled data, which makes it more situational. Also, Auto-Encoders (AE) can be used as a dimensionality reduction technique. The difference between AE and other dimensionality reduction techniques is that AE uses non-linear transformations to project data from a high dimension to a lower one. One of the strengths of AE is that they are neural networks, which means they perform well for certain types of data, such as image and audio data. But they require more data to train.

4.2 Accuracy Metrics

The performance of each machine learning classifier is evaluated according to its accuracy. Accuracy is the ratio of properly classified samples to the total number of samples and is widely used by researchers to describe the performance of an Intrusion Detection System (IDS). The accuracy is calculated by the following criteria:

$$Accuracy = \frac{TP + TN}{TP + FP + FN + TN} \tag{3}$$

TP: number of correctly classified normal connections.
FP: number of normal connections wrongly classified as abnormal connections.
FN: number of anomaly connections wrongly classified as normal ones.
TN: number of anomaly connections correctly classified.

5 Experimental Results

In this section, we present experimental results of different classification algorithms conducted by researchers. All classifiers used the NSL-KDD as the dataset. The Table 2 shows the design studies of the intrusion detection system. Table 2 lists the machine learning algorithm used, the author of the publication, the accuracy, false alarm rate and the time taken to build the model in seconds. Based on the above Table 2, the highest accuracy is 99,99% and the lowest is 75.49%. Actually, the highest accuracy belongs to the hybrid classifier, especially

Table 2. Performance comparisons of several approaches for intrusion detection.

Authors	Classifiers	Data preprocessing	Accuracy	False alarm	Time
Subhy et al. [26]	SOM	FS/Transf./Min-max Norma.	75.49	**	**
Sumaiya et al. [27]	Multi-class SVM	Z-score Norma./Chi-square FS	95.492	0.13	10325
Tahir et al. [21]	K-Means/SVM	Trans/FS using CfsSub -setEval with BestFirst search algo/Min-max Norma.	96.01	3.7	**
Manjula et al. [8]	RF	Transf.	99.8	**	**
Albaayti et al. [5]	BN+RT+RF	**	99.99	0.001	24.97
Çavuşoğlu [33]	RF+J48+NB+KNN	Trans./Min-max Norma./Divide data according to protocol /CfsSubsetEval +WrapperSubSetEval FS	99.87	0.001	10.28

SOM: Self Organization Map-**RF:** Random Forest-**BN:** Bayesian Network-**RT:** Random Tree-**Transf.:** Transformation of characters to a numerical value-**Norma.:** Normalization-**FS:** Feature Selection
**indicates data not provided by the author in the paper

Bayesian Network combined with Random Tree and Random Forest [5], with an intrusion detection rate of 99,99% and low false alarm rate of 0.001% but it's slower to build the model with 24,97 s. Hence, we can say that the proposed system by Çavuşoğlu et al. [33] detects attacks better than other intrusion detection systems in the literature and with a low model building time of 10,28 s. It should also be noted that a hybrid classifier with data pre-processing may perform better on a given set of attack data rather than on individual attacks.

6   Deep Learning Overview

Deep learning is a sub-domain of machine learning that uses algorithms inspired by the structure and function of the brain's neural network. While traditional machine learning algorithms are linear, deep learning algorithms are stacked in an architecture of increasing complexity and abstraction. Furthermore, depending on their approach, deep learning can be classified into three subgroups: generative or unsupervised learning, discriminative or supervised learning, and hybrid learning.

6.1   Unsupervised Learning

In the unsupervised learning process, the main purpose is to generate labelled data using unlabelled data. This occurs when the model learns and makes inferences from unlabelled data. There are several methods classified as unsupervised learning, as follows:

- *Boltzman Machine (BM):* BM is a stochastic model that consists of learning a probability distribution from an original data set. It consists of an input layer (visible layer) and one or more hidden layers. A graphical model consisting of stacked layers of BM is called Deep Boltzman Machine (DBM). In the meantime, due to the complexity of BM, the Restricted Boltzman Machine (RBM) was developed. This is a custom BM without any connection between layers, i.e. the units of a single layer are not connected. It learns patterns and extracts important features from the data by reconstructing the input. If several RBM layers are stacked, it is called a Deep Belief Network (DBN). It is used for classification and also as a feature extraction method for dimensionality reduction.
- *Recurrent Neural Network (RNN):* RNN is an artificial neural network (ANN) designed for time series sequences or data. RNN can be used for supervised or unsupervised learning. It can be applied to image subtitling, connected handwriting recognition or speech recognition.
- *Auto Encoder (AE):* AE is a network of artificial neurons designed to recreate the given input. It is used for feature extraction and dimensionality reduction. AE uses back-propagation and deterministic approaches for the learning process to minimize the discrepancy between the original data and its reconstruction. Several layers of sparse auto-encoders in which the outputs of each layer are wired to the inputs of the next layer; called Stacked Auto-Encoders (SAE). In addition, the Denoising Auto-Encoder (DAE) is designed to reconstruct corrupted data.

6.2   Supervised Learning

Supervised learning is learning in which we teach or train the machine using data that is properly labelled, which means that some data is already labelled with the correct answer. An example of supervised learning is the Convolutional Neural Network (CNN), which is suitable for signal and image processing. The CNN is exceptional not only for feature searching but also for feature engineering, which is the process of extracting useful models from the input to help the predictive model understand the model.

6.3   Hybrid

Many Artificial Intelligence fields (AI) adopt hybrid deep architecture which combines both discriminative and generative architectures. Hybrid architecture allows high-level features to be obtained from raw data for a good representation of input data based on statistical learning. An example of hybrid architecture is the deep neural network (DNN). The DNN is based on a multi-layer Perceptron (MLP) with fully connected cascaded hidden layers.

Yin et al. [30] introduced an effective deep learning approach using the Recurrent Neural Network (RNN-IDS). He compared RNN-IDS with other machine learning classifiers presented by researchers such as J48, Support Vector Machine, Random Forest, Artificial Neural Network... The experimental results show that

RNN helps in improving the accuracy of a classifier to achieve effective intrusion detection and performs better than classical machine learning classification methods in both multi-class and binary classification. The proposed classifier achieved a higher accuracy with a low false positive rate, especially for multi-class classification using the NSL-KDD dataset.

The experiment produced an accuracy of 83.28% in KDDTest$^+$ and 68.55% in KDDTest^{-21} for binary classification with 80 hidden nodes and the learning rate was equal to 0.1. For the five-category classification, the accuracy for KDDTest$^+$ and KDDTest^{-21} was 81.29% and 64.67% respectively, with 80 hidden nodes and the learning rate was 0.5. These results are better than those obtained using J48, Multi-layer Perceptron, Random Forest, Naïve Bayesian, Support Vector Machine, and other classification algorithms. But this proposed model spends more time on training.

In this paper [13] a deep learning-based approach for network intrusion detection using Deep Auto-Encoder (DAE) is implemented. DAE can learn a set of features with better ability to classify minority and majority classes to address data imbalance. The DAE-based IDS achieved an accuracy of 96.53% in binary classification and 94.71% in multi-classification on the total 10% of the KDD-CUP99 test dataset. In addition, the DAE-IDS produced a false alarm (FA) of 0.35% in binary classification.

A work proposed by Al-Quatf et al. [4], Self-Taught Learning (STL)-IDS, using the STL framework that is formed by combining the Sparse Auto-Encoder (SAE) and the Support Vector Machine (SVM). Training and testing time is reduced and the accuracy of Support Vector Machine (SVM) predictions of attacks is improved. The model was compared to other classification methods using the NSL-KDD dataset, such as J48, Naïve Bayesian, Random Forest, and SVM. SAE was used for feature learning and dimensionality reduction to produce good performance by improving the accuracy of SVM classification and reducing algorithm training and test times. For the KDDTrain$^+$ (NSL-KDD) dataset, the proposed approach provides an accuracy of 99.416% and 99.396%, respectively, for the two and five-category classifications.

7   Conclusion

In this article, we highlighted the intrusion detection system's importance and compared the performance of some detection classifiers using data pre-processing to the NSL-KDD dataset in order to make intelligent decisions to detect network attacks. Hybrid smart decision technologies using a combination of classifiers, with the addition of supervised or unsupervised data filtering methods, produce an accurate attack detection rate with a reduced false alarm rate. The overall performance of the resulting model is therefore improved.

The main idea is that it is favourable to apply the dimensionality of the data using one or more of the above techniques and then to apply one or more classifiers according to each type of attack.

The presented research will play a major role for future work in combating against the most complex attacks. This state-of-the-art gives in-depth knowledge of the present IDS challenges. It will be a great help for developing IDS technology and deploying Intrusion Detection Systems in an organization. Future work will consist to put forward a novel approach that combines several types of intrusion detection techniques whose main role is to reduce the rate of false positives/negatives. This approach consists of developing reliable, efficient, and secure distributed IDS based on intelligent agent technology to identify and prevent new and complex malicious attacks in this environment. This approach will consider each detection tool as a Reactive Decisional Agent [24] that cooperate to achieve the desired functionality.

References

1. Datasets | Research | Canadian Institute for Cybersecurity | UNB. https://www.unb.ca/cic/datasets/index.html
2. Agre, G., Peev, S.: On supervised and unsupervised discretization. Cybern. Inf. Technol. **2**(2), 43–57 (2002)
3. Al-Qatf, M., Lasheng, Y., Al-Habib, M., Al-Sabahi, K.: Deep learning approach combining sparse autoencoder with SVM for network intrusion detection. IEEE Access **6**, 52843–52856 (2018)
4. Al-Qatf, M., Lasheng, Y., Al-Habib, M., Al-Sabahi, K.: Deep learning approach combining sparse autoencoder with SVM for network intrusion detection. IEEE Access **6**, 52843–52856 (2018). https://doi.org/10.1109/ACCESS.2018.2869577
5. Albayati, M., Issac, B.: Analysis of intelligent classifiers and enhancing the detection accuracy for intrusion detection system. Int. J. Comput. Intell. Syst. **8**(5), 841–853 (2015). https://doi.org/10.1080/18756891.2015.1084705. http://www.atlantis-press.com/php/paper-details.php?id=25868634
6. Amudha, P., Abdul Rauf, H.: Performance analysis of data mining approaches in intrusion detection. In: 2011 International Conference on Process Automation, Control and Computing, Coimbatore, Tamilnadu, India, pp. 1–6. IEEE, July 2011. https://doi.org/10.1109/PACC.2011.5978878. http://ieeexplore.ieee.org/document/5978878/
7. Bala, R., Nagpal, R.: A review on KDD Cup99 and NSL-KDD dataset. Int. J. Adv. Res. Comput. Sci. **10**(2), 64 (2019)
8. Belavagi, M.C., Muniyal, B.: Performance evaluation of supervised machine learning algorithms for intrusion detection. Procedia Comput. Sci. **89**, 117–123 (2016). https://doi.org/10.1016/j.procs.2016.06.016. https://linkinghub.elsevier.com/retrieve/pii/S187770509163081X
9. Bolon-Canedo, V., Sanchez-Marono, N., Alonso-Betanzos, A.: A combination of discretization and filter methods for improving classification performance in KDD Cup 99 dataset. In: 2009 International Joint Conference on Neural Networks, pp. 359–366. IEEE (2009)
10. Boulle, M.: Khiops: a statistical discretization method of continuous attributes. Mach. Learn. **55**(1), 53–69 (2004). https://doi.org/10.1023/B:MACH.0000019804.29836.05. http://link.springer.com/10.1023/B:MACH.0000019804.29836.05
11. Dougherty, J., Kohavi, R., Sahami, M.: Supervised and unsupervised discretization of continuous features. In: Machine Learning Proceedings 1995, pp. 194–202. Elsevier (1995). https://doi.org/10.1016/B978-1-55860-377-6.50032-3. https://linkinghub.elsevier.com/retrieve/pii/B9781558603776500323

12. Eesa, A.S., Orman, Z., Brifcani, A.M.A.: A novel feature-selection approach based on the cuttlefish optimization algorithm for intrusion detection systems. Expert Syst. Appl. **42**(5), 2670–2679 (2015). https://doi.org/10.1016/j.eswa.2014.11.009. http://www.sciencedirect.com/science/article/pii/S0957417414006952
13. Farahnakian, F., Heikkonen, J.: A deep auto-encoder based approach for intrusion detection system. In: 2018 20th International Conference on Advanced Communication Technology (ICACT), Chuncheon-si Gangwon-do, South Korea, pp. 178–183. IEEE, February 2018. https://doi.org/10.23919/ICACT.2018.8323688. https://ieeexplore.ieee.org/document/8323688/
14. Fayyad, U., Irani, K.: Multi-interval discretization of continuous-valued attributes for classification learning (1993)
15. Kohavi, R.: Scaling up the accuracy of Naive-Bayes classifiers: a decision-tree hybrid. In: KDD, September 1997
16. Kohavi, R., John, G.H.: Wrappers for feature subset selection. Artif. Intell. **97**(1), 273–324 (1997). https://doi.org/10.1016/S0004-3702(97)00043-X. http://www.sciencedirect.com/science/article/pii/S000437029700043X
17. Kohavi, R., Sahami, M.: Error-based and entropy-based discretization of continuous features. In: KDD (1996)
18. Kotsiantis, S., Kanellopoulos, D.: Discretization techniques: a recent survey. GESTS Int. Trans. Comput. Sci. Eng. **32**, 47–58 (2005)
19. Kumar, M., Hanumanthappa, M., Kumar, T.V.S.: Intrusion Detection System using decision tree algorithm. In: 2012 IEEE 14th International Conference on Communication Technology, pp. 629–634, November 2012. https://doi.org/10.1109/ICCT.2012.6511281
20. McHugh, J.: Testing intrusion detection systems: a critique of the 1998 and 1999 DARPA intrusion detection system evaluations as performed by Lincoln Laboratory. ACM Trans. Inf. Syst. Secur. **3**, 262–294 (2000). https://doi.org/10.1145/382912.382923
21. Mohamad Tahir, H., et al.: Hybrid machine learning technique for intrusion detection system (2015)
22. Panda, M., Patra, M.R.: A comparative study of data mining algorithms for network intrusion detection. In: 2008 First International Conference on Emerging Trends in Engineering and Technology, pp. 504–507, July 2008. https://doi.org/10.1109/ICETET.2008.80
23. Reza, M., Miri Rostami, S., Javidan, R.: A hybrid data mining approach for intrusion detection on imbalanced NSL-KDD dataset. Int. J. Adv. Comput. Sci. Appl. **7** (2016). https://doi.org/10.14569/IJACSA.2016.070603
24. Romadi, R., Eddahmani, S., Bounabat, B.: IDS in cloud computing a novel multi-agent specification method. 7 (2005)
25. Shrivas, A.K., Dewangan, A.K.: An ensemble model for classification of attacks with feature selection based on kdd99 and NSL-KDD data set. Int. J. Comput. Appl. **99**(15), 8–13 (2014)
26. Subhy, M., Ibrahim, L.M., Basheer, D.: A comparison study for intrusion database (KDD99, NSL-KDD) based on self organization map (SOM) artificial neural network. J. Eng. Sci. Technol. **8**, 107–119 (2013)
27. Sumaiya Thaseen, I., Aswani Kumar, C.: Intrusion detection model using fusion of chi-square feature selection and multi class SVM. J. King Saud Univ. Comput. Inf. Sci. **29**(4), 462–472 (2017). https://doi.org/10.1016/j.jksuci.2015.12.004. https://linkinghub.elsevier.com/retrieve/pii/S1319157816300076

28. Tavallaee, M., Bagheri, E., Lu, W., Ghorbani, A.A.: A detailed analysis of the KDD CUP 99 data set. In: 2009 IEEE Symposium on Computational Intelligence for Security and Defense Applications, Ottawa, ON, Canada, pp. 1–6. IEEE, July 2009. https://doi.org/10.1109/CISDA.2009.5356528. http://ieeexplore.ieee.org/document/5356528/

29. Yang, Y., Webb, G.I.: Proportional k-interval discretization for Naive-Bayes classifiers. In: De Raedt, L., Flach, P. (eds.) ECML 2001. LNCS (LNAI), vol. 2167, pp. 564–575. Springer, Heidelberg (2001). https://doi.org/10.1007/3-540-44795-4_48

30. Yin, C., Zhu, Y., Fei, J., He, X.: A deep learning approach for intrusion detection using recurrent neural networks. IEEE Access 5, 21954–21961 (2017). https://doi.org/10.1109/ACCESS.2017.2762418

31. Zhang, H., Jiang, L., Su, J.: Hidden Naive Bayes. In: AAAI (2005)

32. Zhang, H., Huang, L., Wu, C.Q., Li, Z.: An effective convolutional neural network based on smote and gaussian mixture model for intrusion detection in imbalanced dataset. Comput. Netw. 177, 107315 (2020)

33. Çavuşoğlu, Ü.: A new hybrid approach for intrusion detection using machine learning methods. Appl. Intell. 49(7), 2735–2761 (2019). https://doi.org/10.1007/s10489-018-01408-x

A Weighted LSTM Deep Learning for Intrusion Detection

Meryem Amar[✉] and Bouabid EL Ouahidi

I.P.S.S, Mohammed V University, Rabat, Morocco
amar.meryem@gmail.com, bouabid.ouahidi@gmail.com

Abstract. The usage of the internet and its opportunities bring not only resources availability, services and storage but puts also customer's privacy at stake. Connected devices share the same pool and Service Level Agreement and are subject to several cyber security challenges. These challenges are either for competitive purposes or to promote the country destruction by warfare attacks.

Deep Learning is robust in easing complex and high dimensionality database analysis to discover relevant predictors. In this paper, we avail the strength of a Weighted Long Short-Term Memory (WLSTM) DN to mine network traffic and prevent the occurrence of attacks. Before attacks identification, the approach pre-processes network traffic using Data Preparation Treatment method to anticipate missing features, and performs after that a weighted conversion on categorical features to discriminate normal behaviors from malicious ones. Afterwards it communicates the weight to the following LSTM classifier. The prevention of attacks is a more challenging task for cyber agent. Thus, Deep belief Network is used to underline joint probabilities over observed traffic and labels.

Keywords: Cyber security · Intrusion detection · Deep learning · Weight of evidence · LSTM

1 Introduction

Nowadays, internet is an unavoidable tool of everyone's daily life, it eventually enhances the number of interconnected networks [1] but brings also several security threats.

The main challenges of each Cyber Security actor are then to prevent these threats before their detection. In order to protect users from unauthorized access, malfunctions and modifications many systems, firewalls and IDSs are configured to block a suspected behavior and raise and alarm when a malicious activity occurs.

Several techniques were proposed to detect malwares that can be divided into two classes:

Rule-based detection (misuse) [2]: this technique compares the input connection to a list of known attack's signatures or malicious behaviors. These rules may be a session number, a source code in an http link (XSS attack) … when the database of known abnormal event is a list of blacklisted connections this is called a *Negative Security model*. Which is simple to process and has very low rate of false positive.

© Springer Nature Switzerland AG 2020
M. Belkasmi et al. (Eds.): ACOSIS 2019, CCIS 1264, pp. 170–179, 2020.
https://doi.org/10.1007/978-3-030-61143-9_14

However, these rules are static and not updated with new attack's signatures. Unlike Negative security model there is also a *Positive Security model* that rejects all connections by defaults except for whitelist ones that are considered as normal. (Firewalls work on Positive Security Model, and the whitelist is done by a cyber security agent or during the learning phase)

Anomaly-based detection uses ML algorithms [3, 4] to collect information at runtime, prepares a model, and then compares input connections to the model. This technique is based on dynamic rules and when the new comportment deviates from the model it is considered as suspicious. These are also called outliers, that may be the number of rejected connection attempts, the number of times the password file is opened, or the size of the uploaded data.

Despite the fact that Machine Learning proposes strong architecture in identifying attacks and especially new malicious behaviors it fails in high rate of false positive when it needs a manual update of selected features. Deep Learning overcomes this limitation and offers a suitable architecture capable of detecting attacks even in a high data dimensionality and complex network traffic.

In this article, we propose a Deep Learning architecture that considers using previous attacks characteristics based on Recurrent Neural Network (RNN). It builds validating patterns depending on the Weight of Evidence of each predictor feature and label several behaviors. Restricted Boltzmann Machine (RBM) layer is added over WLSTM Neural Network to compact the results, minimize the prediction errors between the observations and reconstructed vectors of behaviors. RBM underlines also joint probabilities between traffic behavior and data.

The rest of this paper is divided as follows. Section 2 provides nowadays security challenges. Section 3 depicts a survey of related works. Section 4, details our proposed architecture. To finally provide a conclusion at Sect. 5.

2 Cyber Security Challenges

As cybercriminals perform new attacks and exploit new vulnerabilities, ensuring cybersecurity is becoming tougher. Digitization of factories, development of industrial internet of things, and artificial intelligent technologies make IT networks more connected and communicative. The rise of the number of machines to machine communication gave also rise to more threats and security flaws [5].

Attacks come from different sources: cybercriminals asking for ransom such as WannaCry; stakeholders looking for a competitive advantage or attacker's weakling a country in which industrialist is attacked performing cyberwarfare that is the fifth domain of warfare after land, sea, air, and space [6].

With the complexity and increasing number of cyber-attacks, considerable works are recommended to secure sensitive data from malwares; intrusions [8]; phishing [9, 10]; and spams [11].

3   Literature Survey

Many researches have been done to respond to cyber security issues, one of them were based on supervised algorithms (classical Deep Learning Reinforcement algorithm [7]) and others non supervised (Word Embedding algorithms).

Motivated by the fact that traditional machine learning fails in a high rate of false alarms, Zhendong Wu and Jingjing Wang proposed an algorithm named SRDLM in detection intrusions [12]. That method, starts by re-encoding the semantic of network traffic and remapping it. The goal behind this phase is to represent the network traffic as a stream of character's words in order increase the distance between normal traffic characters and attack's traffic characters and then enhancing the classification accuracy. After that, they have used ResNet Deep Learning algorithm to make its two mapping (residual and identity) stabilize the accuracy when the network deepens.

Jianwu Zhand and Yu Ling [13] proposed a unified model combining multiscale convolutional neural network to analyse the special features of the dataset with long short-term memory to process temporal features and does a classification. The MSCNN-LSTM model had great accuracy, poor false alarm rate and weak negative rates, but does not cover the vanishing gradient problem and requires longer time processing than traditional ML algorithms.

Li, Ma and Jiao [15] proposed a hybrid malicious code detection model based AE and Deep Belief Network (DBN). Where AE is used for features reduction and DBN is the classifier. DBN was trained using a stack of RBMs and BP algorithm to fine-tune the weight and biases of the entire network.

4   Proposed Solution: WLSTM Deep Learning Approach to Detect Attacks

Our proposed architecture is structured as follows (Fig. 1) on three main steps:

1. Data Preprocessing
2. WLSTM Intrusion Detection

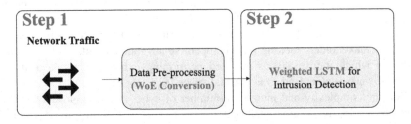

Fig. 1. Proposed architecture

4.1 Data Pre-processing

Network traffic comes from different connected machines and contains different information about connections made but complementary and necessary for attacks' detection.

Network traffic has information about the number of failing attempts, responses of IDS and Firewalls, status of connection at flags (S0, S1, RSTO...) granularity. It records also accessed resources (attempts to access hot files, password file, privileges document...) plus the destination that was targeted.

All of these network traffic connections have non valuable and repetitive information that makes the processing much more time consuming.

Data preprocessing converts network traffic variables into a simpler form that eases upcoming analysis. These variables may be categorical and they go through a "Remapping" phase or "continuous" and they should be discretized into different bins of same length [16, 17]. This is called "data transformation". The second step (see Fig. 2), defines new values of input features using Weight of Evidence (WoE) [18]. This final step, is used to capture the probability of the presence of a feature in an attack or a legitimate behavior.

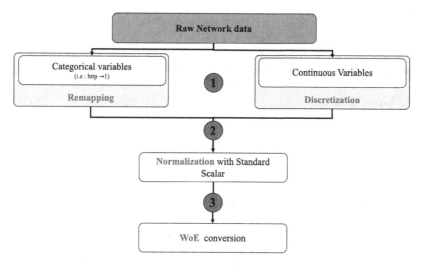

Fig. 2. Phases of the data pre-processing

4.1.1 Step 1: Variable Re-encoding

In this step, we first split network raw data into two categories of variables: categorical and continuous.

For categorical variables (protocol, service...), we are using **Remapping** transformation to get discrete variables. This new values groups customer's behaviors into similar groups. For example, each connection that was under TCP is mapped to connection 1 and for each connection that was by telnet protocol is mapped to connection 7...

For continuous we are using a '**Discretization**'. This transformation, replace the continuous value by a width. The width is obtained first sorting the different values of that variable and calculating the gap between the max and min value. This gap is then divided by the number of bins (b) we which to split data into. For example, if b = 5 then we have 5 epoch and after the 5 times processing data the gradient is forwarded and error rate is reevaluated.

$$\Omega = \frac{x_{max} - x_{min}}{b} \tag{1}$$

4.1.2   Step 2: Normalization

After the categorical and continuous transformations, we pass data through a '**Normalization**'. This step, helps anticipating missing values and thus accelerates the optimization of error (Vanish Gradient problem). We choose to use Standard Scalar transformation:

$$X = \frac{X - \bar{X}}{\sigma} \tag{2}$$

Where $\bar{X}$ is the mean of X and σ is its standard deviation.

4.1.3   Step 3: Weight of Evidence

In order to separate suspicious behaviors from legitimate ones, we are calculating the presence of each variable in a normal event as same as its presence in a suspicious one.

The presence or the absence of this variable can be transcribed by its weight of evidence. Weight of Evidence (WoE) conversion replaces the value of each variable with its strength in labeling either normal behavior or anomalous ones [19]. This method represents then the variable depending on the context [20]. It refers to the strength of a category to separate suspicious behaviors from legitimate ones.

$$WoE_i = \ln\left(\frac{P_{legitimate}(i)}{P_{attacks}(i)}\right) \tag{3}$$

When WoE is near to 0, this means that the probability of feature having the value i to label an unknown behavior as legitimate is near to its probability to label it as normal. Which also means that the value i of that feature is unable to label the connection. Then, it is considered as poor predictor when it has i as value.

After that, we are calculating the information value (IV) for all these features to select the more predictive ones.

$$IV_i = \sum \left(P_{legitimate}(i) - P_{attacks}(i)\right) \times WoE_i \tag{4}$$

Value information (VI) can be interpreted as follows:

- <0.02: The predictor is not useful to label log files events
- [0.02 − 0.1[: The predictor is weak in labeling behaviors

- $[0.1 - 0.3[$: The predictor has a medium strength in labeling attacks
- ≥ 0.3: The predictor has a strong relationship to label the behavior

4.1.4 Data Pre-processing Implementation Results

We have used in our implementation NSLKDD [21] features that we consider as a database of attacks and normal behaviors signatures of traffic network. Based on the results of WoE conversion, we may conclude that: each time we find a (http) protocol we represent it by its corresponding weight (−2.711083) which means that the proportion of legitimate behaviors is larger than the proportion of anomalous ones when the connection is http (Table 1).

Table 1. Some features corresponding WoE conversion

Feature	Min value	Max value	Anomaly rate	Normal rate	WoE
Duration	0	1	0.4834	0.5165	0.0691
duration	2	42862	0.2200	0.7800	−1.1300
protocol_type	Icmp	icmp	0.8422	0.1577	1.8109
protocol_type	Tcp	tcp	0.4796	0.5203	0.0541
protocol_type	Udp	udp	0.1673	0.8326	−1.4686
Service	IRC	IRC	0.0000	1.0000	0.0000
service	Auth	auth	0.7830	0.2169	1.4192
service	Bgp	bgp	1.0000	0.0000	0.0000
service	Courier	courier	1.0000	0.0000	0.0000
service	csnet_ns	csnet_ns	1.0000	0.0000	0.0000
service	Ctf	ctf	1.0000	0.0000	0.0000
service	Daytime	daytime	1.0000	0.0000	0.0000
service	Discard	discard	1.0000	0.0000	0.0000
service	Domain	domain	0.9174	0.0825	2.5435
Service	domain_u	domain_u	0.0016	0.9983	−6.2707
Service	Echo	echo	1.0000	0.0000	0.0000
service	Exec	exec	1.0000	0.0000	0.0000
service	ftp_data	ftp_data	0.2831	0.7168	−0.7932
service	ftp	ftp	0.4898	0.5101	0.0949
service	http	http	0.0548	0.9451	−2.7110
service	http_443	http_443	1.0000	0.0000	0.0000
service	http_8001	http_8001	1.0000	0.0000	0.0000
is_host_login	0	0	0.4661	0.5338	0.0000
is_guest_login	0	0	0.4680	0.5319	0.0074
is_guest_login	1	1	0.2652	0.7347	−0.8834
count	1	4	0.2265	0.7734	−1.0924
count	5	108	0.2911	0.7088	−0.7543
count	109	511	0.8850	0.1149	2.1765
srv_count	1	8	0.4066	0.5933	−0.2424

From Table 2, we understand that the following features are strong predictors:

- Transferred/ downloaded destination bytes feature (dsl_bytes)
- The flag of the connection (Flag, for example: S0, S1...-)
- The corresponding service (Service, for example: TCP...),
- Its protocol type (http, ftp...)

However, we cannot say that the rest of features is not a good predictor but just weaker than the precedent relevant ones.

Table 2. Top 26 of features' information value

Feature	VI	Feature	VI
dst_bytes	4.215066e+00	dst_host_count	1.513907e−02
Flag	4.165250e+00	num_compromised	1.023762e−02
Service	4.067884e+00	is_guest_login	6.567946e−03
src_bytes	3.044355e+00	hot	2.983765e−03
logged_in	2.840624e+00	num_file_creations	2.579018e−03
dst_host_srv_count	2.789469e+00	num_access_files	2.142344e−03
dst_host_same_srv_rate	2.175513e+00	root_shell	1.566713e−03
dst_host_diff_srv_rate	2.013777e+00	num_shells	9.911019e−04
Count	1.759128e+00	wrong_fragment	2.556267e−04
protocol_type	3.929879e−01	num_failed_logins	1.328456e−04
duration	7.770190e−02	land	1.463847e−06
srv_count	6.373327e−02	su_attempted	9.349347e−07
num_root	1.692721e−02	urgent	7.252043e−09

The advantages of replacing discrete variables with only one output variable *WoE* that contains valuable information, avoids creating many additional variables (like in dummy code method) and the values takes into consideration the relationship between the independent and dependent variables. Using a logarithmic transformation makes the distribution of the *WoE* transformed variables a less skewed distribution.

4.2 Weighted LSTM for Intrusion Detection

At this step, we label network traffic to detect intrusions based on NSLKDD features and DL algorithms. We choose DL algorithms in spite of ML ones because ML feature engineering requires a manual feature design when DL extracts the features automatically, and addresses large-scale data problems.

Long Short-Term Memory Networks are special kind of Recurrent Neural Network that works better in training sequential data, such as video capture, word prediction, and image processing. LSTM is composed of three gates (see Fig. 3) and has the characteristic of remembering the elements of a sequence using a memory cell [14]. Then, overcomes the vanishing gradient problem.

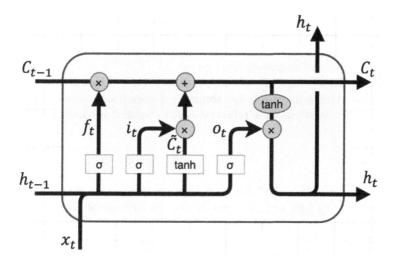

Fig. 3. LSTM cell

Standard RNN are unable to remember information for a very long time because of the side effects of back propagated gradient that either grow or shrink at each step, making the training weights either explode or vanish notably.

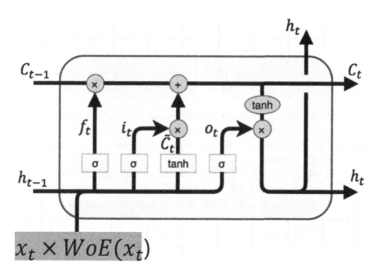

Fig. 4. WLSTM cell

Our weighted LSTM (WLSTM) is composed of three gates that regulates information into and out of the memory cell (see Fig. 4). Instead of using traditional sigmoid layer σ a way to optionally let information through. We are multiplying the input information by its Weight of Evidence (WoE) ($x_t \times WoE(x_t)$ instead of x_t) that gives numbers between zero and one that describes how much each component should be left through depending on its robustness as predictor, as previously calculated in preprocessing step. A value of Zero for a predictor means "let nothing through" and a value of one means "let everything through".

Finally, we consider the results of previous ones in detecting intrusions and update the attack's signatures. The more we update accurately the baseline the best are the results. In the previous work, AUC indicator reflects the great results of that approach [22].

5   Conclusion

In this article, we have proposed a deep learning approach to detect attacks using LSTM deep learning with weighted features through a probability calculation. LSTM gets profile patterns that have information about historical users' behaviors.

Our idea was to define a weight for each feature. Indeed, each value of a feature is multiplied by its weight (WoE). Our implementation seems to prove the idea. In our implementation, we use Tensorflow/Keras with two hidden layers (256 and 128 neurons), sigmoid as the activation function and NSKDD database.

References

1. Miniwatts Marketing Group: World Internet Users Statistics and 2018 World Population Stats (2015). https://www.internetworldstats.com/stats.htm
2. Shah, J.: Understanding and study of intrusion detection systems for various networks and domains. In: 2017 International Conference on Computer Communication and Informatics (ICCCI), pp. 1–6 (2017)
3. Sreeram, I., Vuppala, V.P.K.: HTTP flood attack detection in application layer using machine learning metrics and bio inspired bat algorithm. Appl. Comput. Inform. **15**, 59–66 (2017)
4. Salaken, S.M., Khosravi, A., Nguyen, T., Nahavandi, S.: Extreme learning machine based transfer learning algorithms: a survey. Nerocomputing **267**, 516–524 (2017)
5. Samaneh, M., Ali, G.: Application of deep learning to cybersecurity: a survey. Neurocomputing **347**, 149–176 (2019)
6. Kenneth, K., Vitalice, O., Kibet, L.: Cyber security challenges for IoT-based smart grid networks. Crit. Infrastruct. Prot. **25**, 36–49 (2019)
7. Manuel, L., Belen, C., Antonio, S.: Application of deep reinforcement learning to intrusion detection for supervised problems. Expert Syst. Appl. **141**, 112963 (2020)
8. Markus, R., Sarah, W., Deniz, S., Dieter, L., Hotho, A.: A survey of network-based intrusion detection data sets. Comput. Secur. **86**, 147–167 (2019)
9. Mahmoud, K., Youssef, I.: Phishing detection: a literature survey. IEEE Commun. Surv. Tutor. **15**, 2091–2121 (2013)
10. Douzi, S., Amar, M., El Ouahidi, B.: Advanced phishing filter using autoencoder and denoising autoencoder. In: Proceedings of the International Conference on Big Data and Internet of Thing, pp. 125–129 (2017)
11. Douzi, S., Amar, M., El Ouahidi, B., Laanaya, H.: Towards a new spam filter based on PV-DM (paragraph vector-distributed memory approach). Procedia Comput. Sci. **110**, 486–491 (2017)

12. Wu, Z., Wang, J., Hu, L., Zhang, Z., Wu, H.: A network intrusion detection method based on semantic re-encoding and deep learning. J. Netw. Comput. Appl. (2020)
13. Zhang, J., Ling, Y., Fu, X., Yang X.: Model of the intrusion detection system based on the integration of spatial-temporal features. Comput. Secur. **89**, 101681 (2020)
14. Kim, J., Kim, J., Thu, H.L.T., Kim, H.: Long short term memory recurrent neural network classifier for intrusion detection. In: Proceedings of 2016 International Conference on Platform Technology and Service (PlatCon) (2016)
15. Jin-Young, K., Seok-Jun, B., Sung-Bae, C.: Zero-day malware detection using transferred generative adversarial networks based on deep autoencoders. Inf. Sci. **460–461**, 83–102 (2018)
16. Henrickson, K., Filipe, R., Francisco Camara, P.: Chapter 5 - data preparation. In: Mobility Patterns, Big Data and Transport Analytics, pp. 73–106. Elsevier (2019)
17. Stefan, S., Milad, M., Bjorn, R., Christoph, N.: Social media analytics – challenges in topic discovery, data collection, and data preparation. Int. J. Inf. Manage. **39**, 156–168 (2018)
18. Hosmer Jr., D.W., Lemeshow, S., Sturdivant, R.X.: Applied Logistic Regression. John Wiley, New York (2013)
19. Julie, E.G., KeZu, C.: Short-term ozone exposure and asthma severity: weight-of-evidence analysis. Environ. Res. **160**, 391–397 (2018)
20. Siddiqi, N.: Credit Risk Scorecards: Developing and Implementing Intelligent Credit Scoring. Wiley, Hoboken (2012)
21. Aggarwal, P.: Sharma, S.K.: Analysis of KDD dataset attributes-class wise for intrusion detection. Procedia Comput. Sci. **57**, 842–851 (2015)
22. Amar, M., EL Ouahidi, B.: Hybrid intrusion detection system using machine learning. Netw. Secur. **2020**(5), 8–19 (2020)

Proposed Solution for HID Fileless Ransomware Using Machine Learning

Mohamed Amine Kerrich[1], Adnane Addaim[1(✉)] ⓘ,
and Loubna Damej[2]

[1] ENSA, Ibn Tofail University, Kenitra, Morocco
kerrichamine@gmail.com, adnane.addaime@uit.ac.ma
[2] AB Conseils Information-Technology, Casablanca, Morocco
damejloubna@gmail.com

Abstract. Ransomware is a malware category that asks for payment usually in crypto-currency like Bitcoin after encrypting the files of infected computers. In today's digital threat environment, the rate of ransomware infection has trended upwards to dominate the cyber threat landscape and become one of the emerging threats facing organizations and individuals. In addition to the traditional ransomware, new types have become predominant and they are called fileless ransomware. They use legitimate administration programs and turn them into a tool for fileless infection which is undetectable with traditional security solutions that are based essentially on signature-based detection. In addition to this fileless characteristic, this kind of threat can also be combined with USB-based attacks, also known as HID (Human interface device) attack. In this article, we will present a new attack vector that combines, HID attack, Networking, and fileless ransomware attack on Windows Operating System (OS) machine and then we will present a proposed solution that involves the logistic regression as a Machine learning technique to mitigate against the HID Fileless Ransomware attack.

Keywords: Ransomware · Fileless-ransomware · Signature-based detection · HID attack · Machine learning · Logistic regression

1 Introduction

Recently, the security landscape has witnessed a growing variety of ransomware attacks. This message "All files on your computer have been encrypted. You must pay this ransom within 72 h to regain access to your data." shows that your computer has fallen victim to ransomware, and announces that the attacker has restricted the access to an individual's or enterprise's crucial information and being held hostage and demand a payment in bitcoins to restore the computer to its prior state. The main threat posed by ransomware has varied from disruption, reputation damage, IT resources destruction to financial losses that were estimated in the year 2018 to $1.8 billion with the spreading of ransomware variations like CryptoLocker, GoldenEye, Locky, WannaCry [1, 2]. This extortion exploits the victim's fear about losing vital data, disclosing confidential information or sealing key assets.

© Springer Nature Switzerland AG 2020
M. Belkasmi et al. (Eds.): ACOSIS 2019, CCIS 1264, pp. 180–192, 2020.
https://doi.org/10.1007/978-3-030-61143-9_15

Throughout history, ransomware was unattractive to many adversaries for many reasons such as the low popularity of laptops, the inaccessibility of public web services, the access limitation to reliable cryptographic techniques and thus the lack of untraceable means of payment.

Other types of ransomware have emerged since the end of 2018, they are generally called fileless ransomware [3] because their main behavior is to operate without placing malicious executables on the file system. With this new class of threats, traditional solutions that operate on signature-based detection won't be efficient and are easily bypassed [4], especially if it's combined with a USB based attack [5, 6]. The organization of this paper is as follow: Sect. 2 will give an overview of how ransomware is a tool for cyber extortion, its propagation and entry points and also its cryptocurrencies dependencies, Sect. 3 will discuss Signature-based detection and why it is no more effective with a fileless threat and how it can be bypassed. The common fileless attack vector will be presented in Sect. 4. In Sect. 5 and 6, we will present related works for HID fileless ransomware and our proposed solution respectively. Section 8 will conclude the paper.

2 Ransomware as a Tool of Cyber Extortion

Ransomware is a category of malware that attacks and seizes user-related assets and enables cyber extortion for financial gain [7]. Hackers can send disguised attached files to ransomware as recognizable emails or web pages. Once executed, ransomware prevents victims from interacting with their IT assets, until the extortion is paid out, usually by using a digital crypto-currency such as Bitcoin [8].

These attacks can't only lock down employees who need to get information, but also can utilize shared files as a trigger mechanism to infect other computers in order to quickly spread malware. In 2016, the FBI reported that ransomware caused a total of US\$ 360 million in monetary losses for U.S citizens [9] with a ransom price of around \$300. Also, healthcare facilities were attacked when hackers broke into the system and disabled doctors and nurses from their patients' computerized records, in California's Hollywood Presbyterian Medical Center, and claimed a ransom of about \$17,000 in Bitcoins [10].

Over this course, another attack spread in the manufacturing sector [11] taking advantage of the lack of prevention measures through unrefined security policies, misconfigured systems and the staff un-awareness. In 2017, ransomware continued to experience record growth with the WannaCry outbreak in May 2017 infecting more than 230,000 Windows PCs worldwide using SMBv1 protocol due to the late update process dating back to 2013 left the Microsoft OSs potentially exposed to the most devious hackers, resulting in many victims on government agencies and hospitals [12].

Like WannaCry, Petya was the second major global ransomware in two months using EternalBlue exploit, inflicting a serious disruption at large firms, and infecting systems in Spain, Germany, Israel, the UK, Netherlands and the US, where the government, banks, state power utility, airport, and metro system were all affected. In the same way, a message was displaying and demanding a Bitcoin ransom worth of \$300 [9, 13].

Today's ransomware attacks have focused on industries that have no choice but to pay, such as healthcare, small and medium-sized businesses (SMBs), critical infrastructure, governments and education. Since those attacks, ransomware has moved into different levels of financial payment (see Fig. 1) and data extortion techniques.

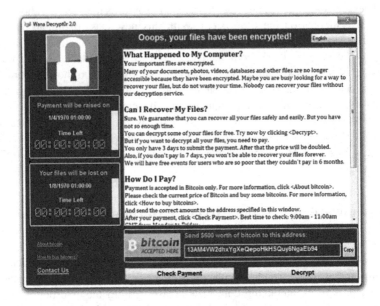

Fig. 1. Payment demand from the WannaCry ransomware

2.1 Ransomware Propagation and It's Entry Point

According to cybersecurity reports, ransomware could be propagated through several malware distribution strategies (see Fig. 2). Some of these can be social engineering, exploit kits, phishing e-mails, weaponized websites, infected attachments, malicious hyperlinks, or simply through a drive-by-download attack or other methods.

The malicious code infection often comes from legitimate sources that are sophistically designed to convince the target to execute the malware without having the least suspicion of its content. We cite below some malware distribution strategies [14]:

- **Phishing e-mails**: It is a fraudulent practice of sending e-mails linking to hosted malware in order to distribute.
- **Social engineering attack:** It exploits human vulnerability rather than software vulnerability. The attack process begins with information gathering, relationship development, and exploitation. In fact, the hacker is going through this malware distribution strategy that plays a vital rule in the process of persuading a victim to open a malicious executable or website that allows ransomware to get a foothold on the victim's system and pressuring the victim into paying for data recovery.

- **Drive-by-download:** This method is used by cybercriminals to compromise a website and inject or embed malicious objects to be downloaded and executed in the background without the user's knowledge.
- **Embedded hyperlinks:** Systematically, these are links contained within documents such as office documents or in Skype instant messaging or any other types of application Do not use the word "essentially" to mean approximately or effectively containing embedded macro viruses that expect users to click on it to download ransomware and install it.

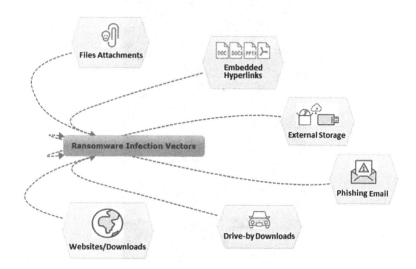

Fig. 2. Ransomware distribution strategies

2.2 Cryptocurrencies and Ransomware Attacks

Bitcoin is the digital currency of choice for cyber extortion which offers a secure and untraceable method of making and receiving money. The invention of Bitcoin [15] was established in 2008 by Satoshi Nakamoto and it was launched in January 2009 when he first began working on a purely peer-to-peer version of electronic payments that could allow online transactions to be sent from one person to another without passing via a bank. The electronic payment system is based on cryptographic proof. In fact, Bitcoin owners don't need any central bank or authority to prove they have funds, and transactions on the Bitcoin network are pseudonymous which makes it impossible to link the Bitcoin account addresses to the real-world identities, also it is not location-based. So, currency can be sent across borders without any limitations. These characteristics make Bitcoin very attractive to ransomware developers as a payment method for their schemes.

In fact, a BlockChain Protocol [16] is a shared, trusted and public ledger of transactions, and it is considered as the technology behind Bitcoin. It operates on top of the internet on a P2P network of computers. The protocol uses a combination of three technologies: cryptography, P2P network, and Game theory. Cryptography relies on the use of public-key cryptography and cryptographic hash functions to ensure transparency and privacy. The P2P networks are used in such a way every node of the network is a client as well as the server. Both are holding identical copies of the application state and finally, with Game theory, the nodes of the P2P network try to validate transactions by consensus algorithm such as proof of work. The solution to the proof of work for the block is broadcasted through the network (Fig. 3).

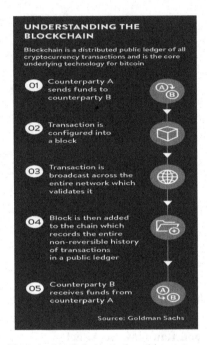

Fig. 3. How the blockchain works with cryptocurrencies.

3   Signature Based Detection

Signature-based detection [4, 17, 18] is the most fundamental method for detecting malicious executable. It's a technique that consists of comparing executable's hashes with a dataset of hashes known threat which means that this solution detects only previously identified malicious executables. Once a new threat has emerged the executable signature is stored in a viral database that would be then checked by the antivirus every time an executable has to be analyzed. This signature that identifies each threat is extracted from different executable sections. These sections represent the necessary information that is used by the operating system to load the executable.

In the windows operating system, a Portable Executable (PE) file is the file format for the executable that contains, in addition to the header section, different other sections that are used to extract the signature.

Below is a list of information and sections parts of the PE file that signature-based detection solutions use [19]:

DLLs: Dynamics Link Library is a type of PE file that stores and regroups multiple Windows API features based on the type of operational tasks that can be performed. For example, crypt32.dll provide function API for Certificate and Cryptographic Messaging functions.

Windows API Functions: All the windows API functions are present within the PE file and can be extracted, Microsoft provides a very well documentation of its API functions under MSDN.

ASCII Strings: ASCII strings within a PE file can give you an insight about the behavior of the file, in a malicious context, if the program creates a file, the filename is stored as a string in the binary, if there is a name resolution with a C2 server controlled by the attacker the domain name will also be stored as string in the binary.

Entropy [20]: It's the amount of data that is presented in a selected file; it can be calculated through the ".resources" section of a PE file that contains all the different data that will be processed by the software at the execution time. High entropy of a PE file means that it contains compressed or encrypted data. We can see that protection's solutions based on this analysis have some limitations and there is no approach with the program's behavior, in a fileless context it's not an effective solution.

This kind of analysis can also be bypassed in malware or no fileless attack with obfuscation [21] which is a technique that attackers use to purposefully obscure their source code. Its main goal is to alter the human readability of a code and its simplicity. Obfuscation can also be used in the context of API calls where functions can be invoked dynamically at execution time and this is achieved by importing the DLL that contains the API function and then extracting a function handle with some functions that are legitimate and used by all windows programs.

4 Fileless Attack Vector

We can classify a fileless attack vector through its entry point which shows how a fileless attack can happen. A fileless attack vector can be done via drive-by-download exploit (such as the eternal blue vulnerability) or Network-based (such as Wanacry) or compromised hardware (such as the BadUSB attack). In the following, we explain in some detail, these three entry points:

Drive-by-Download Exploit Based [22]: Fileless is a misnomer. In fact, in the context of fileless attack, an initial file may exploit a vulnerability within the operating system and can then execute a shellcode and deliver the payload in memory. The payload, in this case, is fileless but the initial entry vector is a file. The hypervisor-based attack can be included in this category that takes advantage of vulnerabilities in the program used to allow multiple operating systems to share a single hardware processor.

Network-Based [23]: With a network communication that takes advantage of the vulnerability within the target machine, an attacker can remotely have code execution. For example, we cite the famous Wanacry that deliver backdoor within kernel memory by exploiting a vulnerability in the SMB protocol.

Hardware-Based [24]: The Hardware attack vector is actually very wide and includes: Device-based, CPU-based, USB-based and BIOS-based.

- **Device-based:** Infecting the firmware which is the software running on the chipset of a device can lead us into a dangerous fileless attack vector.
- **BIOS-based:** A BIOS is a firmware that runs within a chipset. When a machine is booted up, it initializes the hardware and then transfers control to the boot sector. The reprogrammability of this firmware would lead to a dangerous fileless attack vector.
- **CPU-based:** Modern CPUs are highly complicated and can include components that run firmware. Such firmware may enable malicious code to be executed from the CPU. In 2017, two researchers revealed a vulnerability that could allow attackers to perform code on any modern Intel CPU within the Management Engine (ME) which leads again into a very nearly undetectable fileless attack vector.
- **USB-based [25]:** USB-based attack is also known as HID (Human interface device) attack. All types of USB devices can be reprogrammed with malicious firmware that can interact nefariously with the operating system. This is the case with the BadUSB method, which was demonstrated a few years ago, enabling a reprogrammed USB stick to behave like a keyboard that sends commands via keystrokes to computers that can redirect traffic.

5 Related Work for Hid Attacks

In this section, we present an overview of how wide are HID attack vector and some significant works in the detection of malicious HID attack and we conclude this section with some ransomware prevention works.

Recently, numerous forms of USB attacks have been introduced with the programmability of USB devices. Nissim et al. [26] have exposed 29 types of USB attacks that can be grouped into four classes: reprogrammed USB devices, programmable microcontrollers, electric USB devices, and USB devices without reprogramming. As we can deduce from their study, HID-based attacks are predominantly USB-based attacks. We cite below some of them:

Rubber Ducky [27]: This is a commercial keystroke attack platform. It supports a simple scripting language for writing payloads. Once the USB Rubber Ducky is connected to a computer, it proceeds to pretend to be a keyboard and execute the attack.

PHUKD (Programmable HID USB Keyboard/Mouse Dongle): This is a penetration tester that uses a Teensy microcontroller. It provided the inspiration for the development of URFUKED (Universal RF USB Keyboard Emulation Device), a radio-frequency version that offers remote transmission of keystrokes.

USBdriveby: It allows the installation of backdoors and replaces DNS settings with USB hardware based on Teensy.

Evilduino: It is a hardware USB Trojan-based on an Arduino board, whose price is cheaper than the Rubber Ducky and Teensy.

BadUSB: This is the most famous HID attack that performs attacks with a reprogramming USB flash drive that was uncovered by Nohl and Lell at the BlackHat 2014 Conference.

In 2017, Security researcher Samy Kamkar has released his customizable USB attack platform developed using a Raspberry Pi Zero device. In order to detect and prevent this HID-based attacks, Jiunn-Chin Wang presented a process event graph solution that is based on analyzing native host event logs and the insertion time of the HID with the events that are occurred in a period of time after insertion with a method of guilt-by-association [28]. Some other research introduced the concept of virtualization within the device in order to differentiate between the trusted USB and untrusted USB devices and these policies are implemented into the system so that he can detect if the USB device is allowed to be connected or not [28].

As a fileless threat solution, many kinds of research don't bother using statical analysis because it is no more relevant to nowadays threats. They use, instead, information extracted during runtime of a threat in a controlled environment [29].

Researchers in the reference [30] were using YARA rules which consist of a description based on some textual and the binary pattern of an executable to detect the four most relevant ransomware categories (WannaCry, Locky, Cerber, and Crypto-Wall). This solution can actually be effective in the file context.

Another work based on the dynamic behavior released by George Cabeau [31] which detects if a malware executed in sandbox environment is malicious or not and this is done by monitoring the behavior and also verifying the changes made to the file system, registry keys and internet activity.

6 HID Attack's Implementation

The programmable device that we used in our experiment to implement an HID attack was based on a raspberry pi zero w [32] which is a tiny circuit board of a computer with a 65 mm × 30 mm × 5 mm dimension, powered by a 1 GHz Broadcom BCM2835 processor and with 512 MB of LPDDR2 SDRAM with 2.4 GHz 802.11n wireless LAN. As an operating system, we worked with Raspbian lite OS.

Our customized attack script was developed with python. First, it will emulate an Ethernet device over USB, and then it will work as DHCP server that will provide its configuration to the target client. A WPAD (Web_Proxy_Auto-Discovery_Protocol) entry which is a method used by clients to locate the URL of a configuration file using DNS discovery methods will be implemented and static routes for the entire IPv4 address space will be set to the destination target, then with a rogue HTTP/SMB/MSSQL/FTP/LDAP authentication server supporting NTLMv1/NTLMv2/LMv, requests for a variety of protocols sent from the target host will be retrieved by the device, forcing authentication and then the NTLMv2 hash will be captured by the device. This hash will then be sent to a server that will use the appropriate wordlist to crack it. The figure below shows a photo that illustrates our experimentation hardware.

Once the hash is cracked, the server will send it to our device which will act as a keyboard and then will enter the password in order to unlock the locked windows machine. This means that even if the targeted machine is locked it can get infected.

A malicious scenario using keystrokes injection is then executed with a Ducky script which will lead into launching powershell.exe and will deploy a fileless ransomware with the amber framework directly into memory (Fig. 4).

Fig. 4. HID' attack implementation

7   Proposed Solution

To overcome the limitation of statical analysis [4] and some dynamics analysis based on the parent relationship of processes, which in this case won't be effective, we introduced our solution based on extracting different information related to keystrokes once an external device was introduced into a machine. These extracted information data will be used to predict if the introduced keystrokes are human nature or not.

Some of the features, that we are going to use, are the time between successive pressed keys, the use of remove key, the syntax errors, and the time between words. Based on these features, our aim is to create a dataset that will help us build a machine learning model [33] that can be able to classify and distinguish between HID script and human being user. The figure below outlines the time between 39 keystrokes pressed with our programmable device.

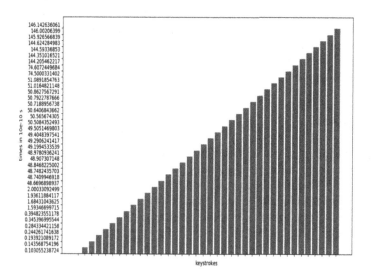

Fig. 5. Time differences between keystrokes

As depicted by Fig. 5, the interarrival time between the pressed key is constant, which does not really reflect the case of a human being behavior that can't perform better than 80 ms between key presses [34].

In our experiments, two target host machines with Windows 10 are configured to collect event log data based on public HID, due to the unavailability of a public HID event log data set. One runs in a virtual machine with 4 GB of RAM and another is on a bare-metal machine with 12 GB of RAM. The virtual machine will record the functionality of the USB-HID scripts and the second machine will collect and record the specific functionality of the human user records.

Depending on our needs, a large dataset of both HID-based scripts and human keystrokes has to be collected. The workflow that we will follow for building our machine learning solution is:

1- **Collect examples of Malicious HID-based scripts, Human keystrokes**. We will use these examples (called training examples) to train the machine learning system to recognize our fileless attack.

2- **Extract** features from each training example to represent the example as an array of numbers. This step also includes research to design good features that will help our machine learning system to make accurate inferences.

3- **Train** the machine learning system to recognize Malicious HID-based scripts using these features.
4- **Test** the approach to some data not included in our training examples to see how well our detection system works.

In terms of algorithms, we have proposed the logistic regression which is a machine learning algorithm that creates a line, plane, or hyperplane (depending on how many features we provide) that geometrically separates our training malicious HID-based scripts from our human keystrokes. In our case, we have common individual features that on their own are strong indicators of maliciousness (small interarrival times of keystrokes, flawless commands...). These features would fit with our logistic regression approach. This approach has proved its efficiency by detecting and preventing from Rubber Ducky and BadUSB fileless attacks.

8   Conclusion

This paper has provided an overview of how the ransomware threats have evolved, how solutions based on statical analyses aren't effective more, and how securing approaches should be. In this paper, we presented a new attack vector combining USB, Networking, and Fileless ransomware attack on Windows operating system, and then we proposed a solution based on machine learning that has detected Rubber Ducky and BadUSB attacks. The ongoing research will test this approach based on logistic regression to many others HID fileless ransomware attacks.

References

1. Carbon B.: Global Threat Report: The Year of the Next-Gen Cyberattack, January 2019
2. Al-rimy, B.A.S., Maarof, M.A., Zainudeen, S., Shaid, M.: Ransomware threat success factors, taxonomy, and countermeasures. A survey and research directions. Comput. Secur. **74**, 144–166 (2018). https://doi.org/10.1016/j.cose.2018.01.001
3. Mansfield-Devine, S.: Fileless attacks: compromising targets without malware. Netw. Secur. **4**, 7–11 (2017). State of Malware Report
4. Moser, A., Kruegel, C., Kirda, E.: Limits of static analysis for malware detection. In: Twenty-Third Annual Computer Security Applications Conference (ACSAC 2007), Miami Beach, FL, pp. 421–430 (2007)
5. Fu, J., Huang, J., Zhang, L.: Curtain: keep your hosts away from USB attacks. In: Information Security. 20th International Conference (ISC 2017), Ho Chi Minh City, Vietnam, 22–24 November, pp. 455–471 (2017). https://doi.org/10.1007/978-3-319-69659-1_25
6. Hernandez, G., Fowze, F., Tian, D., Yavuz, T., Butler, K.R.B.: FirmUSB: vetting USB device firmware using domain informed symbolic execution. In: Proceedings of the 2017 ACM SIGSAC Conference on Computer and Communications Security (CCS 2017), Dallas, TX, USA, 30 October–03 November, pp. 2245–2262 (2017)
7. Tailor, J.P., Patel, A.D.: A comprehensive survey: ransomware attacks prevention, monitoring and damage control. Int. J. Res. Sci. Innov. **IV**, 2321–2705 (2017)

8. Paquet-Clouston, M., Haslhofer, B., Dupont, B.: Ransomware payments in the bitcoin ecosystem. J. Cybersecurity (2018). https://doi.org/10.1093/cybsec/tyz003
9. 2016 internet crime report. Technical report. Internet Crime Complaint Center, Federal Bureau of Investigation (2016)
10. Spence, N., Paul III, D.P., Coustasse, A.: Ransomware in healthcare facilities: the future is now. In: The Academy of Business Research, Fall 2017 Conference. Atlantic City, NJ (2017)
11. Butt, U.J., et al.: Ransomware threat and its impact on SCADA. In: 2019 IEEE 12th International Conference on Global Security, Safety and Sustainability (ICGS3), London, United Kingdom (2019)
12. KSN report: Ransomware in 2014–2016. Technical report, Kaspersky Lab (2016)
13. Symantec: 2018 internet security threat report. vol. 23. Technical report, Symantec (2018)
14. Salahdine, F., Kaabouch, N.: Social engineering attacks: a survey. Future Internet (2019). https://doi.org/10.3390/fi11040089
15. Satoshi, N.: Bitcoin: a peer-to-peer electronic cash system (2008)
16. Sachchidanand, S., Nirmala, S.: Blockchain: future of financial and cyber security. In: 2016 2nd International Conference on Contemporary Computing and Informatics (IC3I), India (2016)
17. Erdogan, O., Cao, P., Hash-AV: fast virus signature scanning by cache-resident filters. In: Proceedings of the International Conference on Systems and Networks Communications (ICSNC) (2007)
18. Nwokedi, I., Mathur, A.P.: A Survey of Malware Detection Techniques (2007)
19. Seungwon, H., Keungi, L., Sangjin, L.: Packed PE file detection for malware forensics. In: 2nd International Conference on Computer Science and its Applications, Jeju, South Korea (2009)
20. Lyda, R., Hamrock, J.: Using entropy analysis to find encrypted and packed malware. Secur. Priv. IEEE 5(2), 40–45 (2007)
21. Canfora, G., et al.: Obfuscation techniques against signature-based detection: a case study. In: 2015 Mobile Systems Technologies Workshop (MST), Milan, Italy (2015)
22. Sood, A.K., Zeadally, S.: Drive-by download attacks: a comparative study. IT Prof. 18(5), 18–25 (2016)
23. Vassilakis, V., Moscholios, I., Logoth, D.M.: Static and dynamic analysis of wannacry ransomware. In: IEICE Information and Communication Technology Forum (ICTF), Austria (2018)
24. Bidmeshki, M., et al.: Hardware-based attacks to compromise the cryptographic security of an election system. In: IEEE 34th International Conference on Computer Design (ICCD), Scottsdale, AZ, USA (2016)
25. Tischer, M., et al.: The danger of USB drives. IEEE Secur. Priv. 15(2), 62–69 (2017)
26. Nissim, N., Yahalom, R., Elovici, Y.: USB-based attacks. Comput. Secur. 70, 675–688 (2017)
27. Cannols, B., Ghafarian, A.: Hacking experiment by using USB rubber ducky scripting. Systemics Cybern. Inform. 15(2) (2017). ISSN 1690-4524
28. Huang, C.-Y., et al.: Identifying HID-based attacks through process event graph using guilt-by-association analysis. In: Published in ICCSP 2019. https://doi.org/10.1145/3309074.3309080
29. Trend Micro: Fileless Malware: A Hidden Threat. Industry News, Security, 23 October 2017
30. Jaramillo, L.E.S.: Malware detection and mitigation techniques: lessons learned from Mirai DDOS attack. J. Inf. Syst. Eng. Manage. 3(3), 19 (2018). https://doi.org/10.20897/jisem/2655

31. Cabau, G., Buhu, M., Oprisa, C.P.: Malware classification based on dynamic behavior. In: 2016 18th International Symposium on Symbolic and Numeric Algorithms for Scientific Computing (SYNASC), Timisoara, Romania (2016)

32. Kurniawan, A.: Getting Started with Raspberry Pi Zero W. PE Press (2017)

33. Torres, P.E.P., Yoo, S.G.: Detecting and neutralizing encrypting Ransomware attacks by using machine-learning techniques: a literature review. Int. J. Appl. Eng. Res. **12**(18), 7902–7911 (2017)

34. Umphress, D., Williams, G.: Identity verification through keyboard characteristics. Int. J. Man-Mach. Stud. **23**(3), 263–273 (1985)

Wireless and Mobile Network Security

Data Oriented Blockchain: Off-Chain Storage with Data Dedicated and Prunable Transactions

Oualid Boumaouche, Afifa Ghenai[(⊠)], and Nadia Zeghib

Lire Laboratory, Constantine 2 – Abdelhamid Mehri University,
Constantine, Algeria
{oualid.boumaouche,afifa.ghenai,
nadia.zeghib}@univ-constantine2.dz

Abstract. Blockchain is a rising technology mainly characterized by its decentralization, persistency, anonymity, and auditability. Due to these properties it is nowadays increasingly used in IoT and smart cities. However, the amount of generated data in these areas is large, and the size of the blockchain increases steadily as transactions and data are stored. To deal with this drawback, the blockchain technology must be adapted, in particular regarding its size. Indeed, after a few years of operation, Bitcoin and Ethereum blockchains already weigh more than 200 GB, which is impossible to manage for resource limited devices. Several solutions have been proposed, each with a different approach. In this paper, we try to overcome many of their limitations, by offering a solution that allows to greatly reduce the size of the blockchain by proposing an architecture specifically dedicated to the storage and manipulation of data, while allowing a free control and sharing of data by their owner.

Keywords: Blockchain · Off-Blockchain storage · Storage optimization · Blockchain size · Digital identity · Blockchain transactions

1 Introduction

Despite the key features of the blockchain (decentralization, persistency, anonymity, and auditability), which make its use very interesting for the Internet of Things (IoT), smart cities and smart governance, these application domains generate a lot of data that needs to be stored. Storing them directly on the blockchain would increase its size in a way that would make it unusable by machines with limited resources (machines used precisely in the fields of IoT, or smart cities, for example). The size of the blockchain increases over time as data, transactions, or smart contracts are stored there. The size of the bitcoin blockchain (headers and transactions) is 229 GB in July 2019 [1]. The size of the Ethereum blockchain, to store for a full node, is 202 GB with the Ethereum Parity implementation, and is 311 GB with the Ethereum Geth implementation, in July 2019 [2]. This is why we must limit the direct storage on the blockchain to the bare minimum and deport the mass of data to be stored outside the blockchain. In addition, solutions must be proposed to purge the blockchain transactions that are no longer relevant. Several solutions have been proposed. In [3], the authors suggest a solution,

© Springer Nature Switzerland AG 2020
M. Belkasmi et al. (Eds.): ACOSIS 2019, CCIS 1264, pp. 195–204, 2020.
https://doi.org/10.1007/978-3-030-61143-9_16

where transactions, after a while, can be forgotten by the network, which limits the size of the blockchain to the most recent blocks. On another side [4] proposes to store the data out of the blockchain in Distributed Hashing Table. Finally, the authors of the article [5] have used a pruning strategy to remove transactions that are no longer relevant and keep only those that are needed, thus limiting the growth rate of the blockchain.

These solutions still have limitations, including the inability of some to store other things than simple cryptocurrency transactions, impossibility of using smart-contracts or lack of control and sharing capabilities on the stored data.

In this paper, we propose a blockchain oriented towards data storage. We use storage outside the blockchain for large data and smart contracts, and we suggest a set of transactions specifically dedicated to data manipulation, allowing better control over them. The remainder of the paper is organized as follows, Sect. 2 discusses in more detail some related work before highlighting their strengths and weaknesses. In Sect. 3 we give an overview of our proposed solution. In Sect. 4 we present the general architecture and then we shift to an application example. Finally, Sect. 5 concludes the paper and presents ongoing work.

2   Related Work

Before presenting the relevant works dealing with the problem cited above, we recall some basic concepts:

- *Blockchain*: can be defined as a distributed ledger, where the data is stored in a list of chained blocks. Asymmetric cryptography and consensus algorithm ensure security and consistency. The key characteristics of the blockchain are decentralization, persistency, anonymity and auditability [8, 9].
- *Smart Contract*: is a program that is executed on the virtual machine constituted by the blockchain infrastructure, the Ethereum infrastructure for example [11].
- *Transaction:* is data that tells the network that a user A transfers to user B a certain amount of cryptocurrency, that a user U has executed a smart contract S, that a user U requests the creation of data D1, or that a service provider S requests the access to a data D2 [10].

In the following, we present the most relevant approaches related to our work. We describe them in detail and we give a comparison table to highlight the strengths and weaknesses of each one of them.

[3] proposes a cryptocurrency scheme where old transactions can be forgotten by the network. It considers that pruning solutions are complex and limited. It aims to stop the growth of the blockchain and provide a really lightweight scheme. This approach eliminates the need for a full blockchain. The remaining part is called "mini-blockchain" and it is composed by the newest blocks of the blockchain. The loss of security this cutting process incurs is solved with a small "proof chain", which consists of a chain of all blockchain headers. The loss of coin ownership data is solved with a database which holds the balance of all non-empty addresses, the "account tree", a hash tree structure. Thus, we don't need transactions to calculate the balance of any given

address. The proof chain secures the mini-blockchain and the mini-blockchain secures the account tree. The scheme allows the transactions to be discarded, after enough time has passed, by unlinking them. It removes the script system and along with it the idea of interlocking transactions. It replaces it with a simpler concept where transactions perform basic operations on the account tree, such as subtracting coins from the balance of address A1 and add to the balance of address A2. Transaction inputs and outputs don't point to other transactions, they simply point to addresses in the account tree. While using the mini-blockchain concept reduces and limits the size of the blockchain. The proof chain continues to grow over time with new block creations; and to a lesser extent the account tree too, with the creation of new addresses. These two last data structures can generate the same origin storage problem in the long run. Another drawback of the mini-blockchain solution is that it is limited to a basic form of transactions. It eliminates the use of smart contracts that allow for more elaborated operations, thus limiting the usage of the solution for cryptocurrencies. The possibility of discarding old blocks prevents the use of blocks for storing other data than transactions.

The authors of the article [4] have proposed a personal data management platform that combines blockchain and off-blockchain storage. It adds two new types of transactions: Taccess used to manage access control to the stored data; and Tdata used for data storage and retrieval. A service can use this platform to store data related to a user. The encrypted data to store is transmitted to the blockchain along with associated access control configuration, in a Tdata and Taccess transactions respectively. The blockchain routes the data into an off-blockchain key-value store, and retains only a pointer (a SHA-256 hash for the data) to the data on the public ledger. The user and the service can, from now on, query the data using a Tdata transaction with the pointer associated to it. The user can change the permissions allowed to the service using a Taccess Transaction. The off-blockchain storage consists of an implementation of Kademilia [6], a distributed hash table (DHT) with added persistence using LevelDB [7] and an interface to the blockchain. A network of nodes, possibly different from the blockchain network, is responsible for the maintenance of the DHT, and for the execution of approved read/write transactions. The use of a off-blockchain storage for application data, reduces the size of the blockchain by avoiding the storage of large data directly on the public ledger. However, the blockchain continues to grow as new blocks, containing transactions and data hashes, are created. Thus, although the blockchain size is greatly reduced by this approach, it can still be a problem for resource-limited nodes.

The authors of the article [5] have presented a pruning strategy similar to the disk space reclaiming procedure proposed in the original Bitcoin paper [8] with the difference that transactions of interest can be left unpruned. The proposed strategy relies on an arbitrary predicate function, for transaction selections. These predicate functions can select only transactions cancelling each other out, or are fully superseded by later transactions, to be deleted. The pruning can only reduce the growth rate of the blockchain, not limit its size, since blocks are not deleted. The blockchain continues to grow as blocks are generated. The predicate functions used to select prunable transactions are not generally applicable and are only able to identify transactions that deal with data that has been removed.

With our contribution, we are remediating the impossibility of storing anything other than cryptocurrency trading transactions, and the usage of smart contracts. We allow the owner of the data to be able to share his data with the entities of his choice and even to be able to monetize them. Finally, we limit the speed of the blockchain size growth by allowing the transaction pruning (Table 1).

Table 1. Related work, advantages and limits comparison

Solutions	Advantages	Limits
[3]	• Limits the size of the blockchain	• Limited to a basic form of transactions • Impossibility to store data Permanently on the blockchain • Impossibility to use smart contracts
[4]	• Reduces the size of the blockchain • Reduces the cost of data storage	• The data is only accessible for one service provider
[5]	• Reduce the growth rate of the blockchain • Interesting transactions can be left unpruned	• The predicate functions used to select prunable transactions are not generally applicable

3   Proposed Solution: Data Oriented Blockchain

Data Oriented Blockchain is a blockchain geared towards storing, controlling, and manipulating data. It allows any entity to store its data, control who has access to it, and even monetize it. Data storage is provided by a storage layer, decentralized, and duplicated. To ensure data operations, the Data Oriented Blockchain is equipped with transaction specifically dedicated for this purpose, which we call Data Oriented Blockchain Transactions. The system has a transaction pruning feature which, if it does not limit the size of the blockchain, decreases its growth rate over time. In addition to data, the storage layer stores smart contracts. There are only their hashes that are stored on the blockchain, to ensure their authenticity by checking them before the execution of smart contracts. The Data Oriented Blockchain also supports a cryptocurrency allowing monetized exchanges between entities. This allows an owner, to sell his data, and reward the nodes storing data and those ensuring the security of the blockchain, by verifying and validating transactions.

3.1   Data Oriented Blockchain Transactions

Data Oriented Blockchain Transactions is a set of transactions dedicated to defining data stored via blockchain, data manipulation, data access, and data control.

Data Definition Transactions are used to define the shape of the data to be stored. We can see an example in Fig. 1.

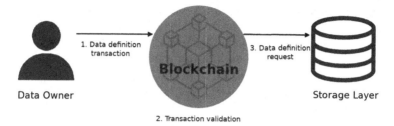

Fig. 1. Data definition transaction example

Data manipulation transactions allow assigning values to data, changing their values, or deleting them. An illustration is provided in Fig. 2.

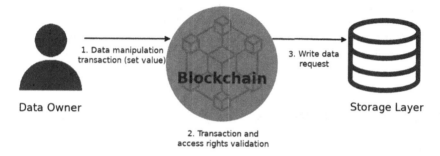

Fig. 2. Data manipulation transaction example

Data access transactions allow different entities to request access to the data. As shown in Fig. 3.

Data control transactions allow the owner to define who may or may not access it, under what conditions, and what manipulations it can apply.

When the user wants to store data, he starts by defining it with Data Definition Transactions, defines who can access it with Data Control Transactions, and finally assigns values to it with Data Manipulation Transactions. When a service provider

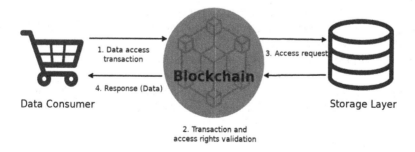

Fig. 3. Data access transaction example

wants to access a user's data, he uses Data Access Transactions. The importance of transactions lies in the fact that they ensure the security of the storage layer by enabling Mining Nodes to verify and validate the operations performed on the storage layer. In addition to this, it allows to keep track of the operations performed and thus ensure auditability.

3.2 Transaction Pruning

The system periodically checks for deletable transactions by selecting them according to pre-established rules. For example, mutually canceling transactions (two consecutive changes in the same data), or transactions applied to deleted data.

A user deletes his data using Data Manipulation Transactions. The Mining Nodes periodically check the existence of transactions eligible for pruning. For example, they discover the transactions relating to the data deleted by the user and delete them.

The rules for selecting transactions are established beforehand at the system level, and may vary according to the objectives to be implemented. If we want to keep the history of all data accesses, even those deleted, we will establish rules that only select the Data Manipulation Transactions for pruning and keep the Data Access Transaction on the blockchain.

4 System Architecture

The system architecture is presented in Fig. 4. First, we present the entities constituting the system, followed by an application example showing their interactions.

4.1 System Entities

We present the entities participating in the system, assuming that data storage is provided by an implementation of the Distributed Hash Table (DHT), where different nodes are responsible for storing data and smart contracts [6]. Nodes involved in transaction validation route data requests to this storage layer.

Data Owner: Identified by a unique identifier calculated from the key pair, private key, public key, as for the Bitcoin system. It is the data owner, the one who creates it by initiating data definition transactions. It controls who has access to it and under what conditions (possibility to put a data access price in the form of cryptocurrency) by using data control transactions. The Data Owner can change the data access control settings at any time, assign or change their value, or delete them using the data manipulation transactions. Data Owner can also initiate the creation of a smart contract which will be identified by a unique address, and stored on storage layer. The Data Owner, can be any entity with an identifier, a person or a service provider.

Data Consumer: Identified by a unique identifier calculated from the pair, private key, public key. It needs to access the data stored by the Data Owner for different reasons. It requests access to data through data access transactions. They are or are not transmitted according to the access rules defined by the Data Owner. Data Consumer can also launch a request to execute a smart contract. The Data Consumer may be a service provider, a government authority, a research organization or any other entity with a unique identifier.

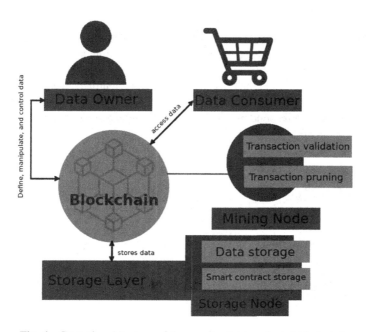

Fig. 4. General architecture of the proposed data oriented blockchain

Mining Nodes: Mining Nodes check the validity of transactions and smart contracts, possibly by checking the data access control settings, specified by the Data Owner. Mining Nodes integrate the validated transactions into a block and begin the actual mining operation. Using a proof of work algorithm in the same way as the Bitcoin system for example. As soon as a node arrives at the mining solution of a block, the node sends the requests corresponding to the transactions of the mined block to the data storage layer, and sends the block to the other Mining Nodes. They check its validity, integrate it into the blockchain, and send the corresponding requests to the data storage layer. In addition to this, the mining nodes take care of the application of the pruning operation.

4.2 Digital Identity Application Example

We use the Data Oriented blockchain to implement the architecture of a digital identity management system that is decentralized and open. Our solution is to store user data on the blockchain and avoid storing it at the service provider. The service provider accesses the data through access transactions. The actors participating in the system are: The user: represents the user interested in using the service. It corresponds to the Data Owner.

Service Provider: Represents the entity providing the service used by the user. He needs information about the user to ensure his service. It corresponds to the Data Consumer.

Use Case Scenario

- Creating the digital identity: Any use of the system starts with the creation of a digital identity. As we can see in Fig. 5, the user, handling an application that uses the Data Oriented Blockchain, creates an account. The application generates a pair of keys, private and public, and calculates a unique identifier, like public addresses of the blockchain. The user enters his personal data. The application then generates the transactions corresponding to this data and sends them to the Data Oriented Blockchain network.

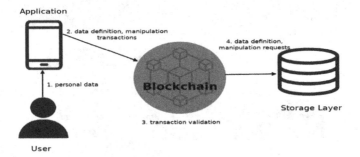

Fig. 5. Digital identity creation

- Validating data creation transactions: As soon as a mining node receives data creation transactions, it validates them and adds them to a block, before starting the mining operation. When the mining is complete and a solution is found, the mining node sends the requests corresponding to the transactions contained in the block to the storage layer, and stores on the blockchain, only the hash of the data as well as the identifier of the Data Owner. The mining node sends the block to the other nodes to validate it and add it to the blockchain.

- Access to a service: As illustrated in Fig. 6, when the user wants to access a service, he provides his identifier to the service provider. The service provider verifies the existence of the identifier on the blockchain, and generates data access transactions. These transactions indicate the data that the service provider wants to access.

Service Provider

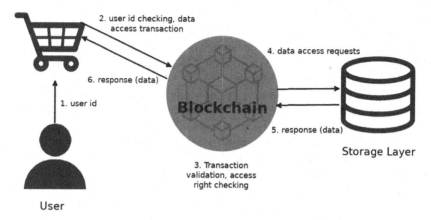

Fig. 6. Accessing service

- Validating data access transactions: When a mining node receives a data access transaction, it checks its validity by accessing the data control parameters specified by the Data Owner. If the Data Consumer is authorized to access this data, it is transmitted to them.

4.3   Smart Contract Application Example

To illustrate the use of smart contracts, we take as an example a transparent payment and donation system, which allows a customer to pay a portion of the amount paid, to a charity. The actors participating in the system are:

e-commerce service: an online merchant offering services or products, donating a part of its income to one or more charities, while accepting payment by cryptocurrency. customer: a person wanting to access the services or products offered by the e-commerce service.

Use Case Scenario

- Creation of the smart contract: to set up its transparent payment and donation system, the e-commerce service writes a smart contract, which accepts as an entry, an amount in cryptocurrency and which pays a share to the account of the e-commerce service. commerce, and another on the account of a charity. The smart contract is sent to the blockchain, checked by the mining nodes and stored on the storage layer. The smart contact's address and hash are kept on the blockchain.

- Payment: To pay for a product or service, the customer, who has the necessary cryptocurrency, launches the execution of the smart contract. The smart contract is then read from the storage layer, and its integrity is verified by the mining nodes, using its hash kept on the blockchain. It is finally executed by the mining nodes.

- The customer can at any time check the content of the smart contract by accessing it and thus ensure that it realizes what the e-commerce service claims.

5 Conclusion

With Data oriented blockchain, we have given the data owners a complete way to manipulate, protect, share or monetize their data, while decreasing the size of the blockchain by storing the data in a dedicated layer, separate from the blockchain. The ability to delete unnecessary transactions further decreases the size of the blockchain by keeping only the necessary transactions.

Further work will consist in the explicit definition of an example of implementation of the rules and protocol of transaction pruning operations, as well as an example of a general application implementing a choice of a storage technology and an implementation of two types of nodes, storage node and mining node.

Despite the big gains in blockchain size, storing data out of it, and deleting transactions as they become obsolete, we are only slowing down its growth rate. A next step would be to limit the size of the blockchain to a maximum value that it will never exceed and that by finding a solution to be able to store relevant transactions, data hashes and data control rules, out of the blockchain, while maintaining the key qualities offered by it.

Another research track, essential to the use of the blockchain in the field of IoT and smart cities, is the optimization in terms of the computing resources required for a node to participate in blockchain security networks.

References

1. Blockchain Homepage. https://www.blockchain.com/learning-portal. Accessed 20 Sept 2020
2. Nodestats Homepage. Blockchain Data products. https://www.nodestats.org/. Accessed 20 Sept 2020
3. Bruce, J. D.: The mini-blockchain scheme. White paper (2014)
4. Zyskind, G., Nathan, O., Pentland, A.: Decentralizing privacy: using blockchain to protect personal data. In: IEEE Security and Privacy Workshops, USA (2015)
5. Palm, E., Schelén, O., Bodin, U.: Selective blockchain transaction pruning and state derivability. In: Crypto Valley Conference on Blockchain Technology, Switzerland (2018)
6. Maymounkov, P., Mazières, D.: Kademlia: a peer-to-peer information system based on the XOR metric. In: Druschel, P., Kaashoek, F., Rowstron, A. (eds.) IPTPS 2002. LNCS, vol. 2429, pp. 53–65. Springer, Heidelberg (2002). https://doi.org/10.1007/3-540-45748-8_5
7. Github Homepage. https://github.com/google/leveldb. Accessed 18 Sept 2020
8. Nakamoto, S.: Bitcoin: a peer-to-peer electronic cash system (2008)
9. Zheng, Z., Xie, S., Dai, H., Chen, X., Wang, H.: An overview of blockchain technology: architecture, consensus, and future trends. In: IEEE International Congress on Big Data. IEEE (2017)
10. Antonopoulos, A.M.: Mastering Bitcoin: Unlocking Digital Cryptocurrencies. O'Reilly Media, Sebastopol (2014)
11. Antonopoulos, A.M., Wood, G.: Mastering Ethereum: Building Smart Contracts and DApps. O'Reilly Media, Sebastopol (2018)

A Framework to Secure Cluster-Header Decision in Wireless Sensor Network Using Blockchain

Hafsa Benaddi[1]([✉])[iD], Khalil Ibrahimi[1][iD], Haytham Dahri[1],
and Abderrahim Benslimane[2][iD]

[1] Faculty of Sciences, MISC Lab, Ibn Tofail University, Kenitra, Morocco
{hafsa.benaddi,ibrahimi.khalil}@uit.ac.ma, haytham.dahri@gmail.com
[2] University of Avignon, CERI/LIA, Avignon, France
abderrahim.benslimane@univ-avignon.fr

Abstract. The Wireless Sensor Network (WSN) is hugely taking a big interest in worldwide in terms of security and private grants. Despite the increase of massive data transmission, a lot of challenges discovered in this technology, the data, and the critical information become vulnerable to a lot of malicious behaviors. In this paper, we proposed a new scheme to develop a dynamical framework to secure cluster-header decision-based blockchain in WSN that aims to facilitate the transactions and efficient to build a low-cost system. However, it applies the concept of Ethereum's smart contracts to verify and check the confidentiality of each node in a decentralized architecture way. Otherwise, the implementation of the framework is deployed on the ganache environment using solidity as a programming language to write smart contracts to reach and resolve both the flexibility and efficiency of proof of work for each node. Consequently, the implementation results of the proposed scenarios confirm that this study achieves integrity, agreement, validity, availability, and termination which make the whole system a promising solution by decreasing building cost and improving the architecture effectiveness and trustworthiness.

Keywords: Blockchain · Wireless Sensor Network · Smart contact · Cryptocurrency · Decentralized control · Mining · Cryptography · Ethereum · Bitcoin

1 Introduction

Up to now, electronic transactions based on *blockchain* have become very popular due to the good reputation of this technology. However, the honor cannot deny the dangerous activities and the risks that can cause these cryptocurrencies [2,14]. In this context, the centralized network of the election systems have been hacked in an illegal way or governments who intimidate their citizens with violence if they don't give their elect for them [6], this happens all the time, but

© Springer Nature Switzerland AG 2020
M. Belkasmi et al. (Eds.): ACOSIS 2019, CCIS 1264, pp. 205–218, 2020.
https://doi.org/10.1007/978-3-030-61143-9_17

blockchain as a novel technology could solve and facilitate the network management dynamically. In fact, blockchain is a shared ledger of transactions between parties in a network, not controlled by a single central authority and distributed over nodes. A ledger is a kind of a record book, it records and stores all transactions between users in chronological order. However, while there are a lot of features, two of the most important are the openness of the platform (public or private).

Firstly, any public blockchain necessarily works with a currency or a token (token) programmable while private blockchain can only be viewed by a chosen group of people. Bitcoin is an example of a public programmable currency, the transactions between network users are grouped in blocks in which each block is validated by the nodes of the network called "miners", according to mechanisms that depend on the type of blockchain. In the bitcoin blockchain, this technique is called the "Proof-of-Work" (PoW) and consists of solving mathematical problems. Once the block is validated, it is time-stamped and added to the blockchain. Consequently, the transaction is then visible to the receiver as well as the entire network. This process takes some time depending on the blockchain we are talking about (ten minutes for bitcoin, 15 s for Ethereum) [11,13]. More precisely, Ethereum is the ultimate and well-established, openhanded decentralized software platform that allows Decentralized Applications (DApps) and smart contracts to construct, build and run without any third party to prevent and exclude any downtime, fraud, control or interception permanently. It uses Ether as a cryptocurrency which is broadly used to run applications, enhance the work and even monetize it. Secondly, they are two levels of permissions required to add the information to the blockchain: the permissioned blockchain permits just a selected group of users to write (i.e. generate transactions for the ledger to record) and commits (i.e. verify new blocks for addition to the chain) [10]. In contrast, permission less Blockchains allow anyone to contribute and add data to the ledger [8]. Thus, Ethereum transaction takes time less than bitcoin.

Therefore, *Wireless Sensor Network* (WSN) is a set of physical sensors that represent a full network that allows and provides a better environment for effective communication. The Sensor node is composed of a radio transceiver connected with an antenna, an interfacing electronic circuit, a microcontroller and a battery as a supplier of energy. One of the most popular architectures of WSN is based on designing a master node to control other devices in the network. Among the existing challenges in sensor networks, we are struggling with the lack of a trusted third party. As a result, a high number of malicious *Internet Of Things* (IoT) devices have been used to create widespread malware. These growing threats the financial area by costing a lot of money of lost revenues for the largest companies [1].

In this paper, the suggested approach can verify the following requirements (reliability, robustness, efficiency, security and ergonomics) which the environment ought to be adapted to the user without providing too much effort. In the technical aspect, we are motivated to provide a newly secured mechanism by taking the advantages of asymmetric encryption, proof-of-work, and transactions of the blockchain network which represent the necessary enhancement for data validation. Even more, our proposal holds the following advantages and features, which can able to be independent of any third party or centralized one and various consensus protocols are needed to validate the data in order to prevent and remove any kind of duplicated entry or fraud because of the smart contracts allow to set pre-conditions based on the business logic.

In this proposed contribution, the process presented as follows:

- We using a private blockchain environment to construct a decentralized and secured peer-to-peer network approach allowing connected users to perform transactions using their permission blockchain's address as an authorization piece.
- We deploy the smart-contract to build a secured system dynamically to choose a *Cluster-Header* (CH) in WSN that can be integrated with the previous model to designate a CH from a set of WSN based-blockchain clusters.
- The simulations results are conducting for evaluating the network transactions when the test scale of the balance of all nodes achieves an already determined value so that the network transactions will be frozen.

The rest of the document is orderly as follows: Sect. 2 offers some backgrounds mentioned in this study. Section 3, focuses on the description of the system process details, characterization, and decentralized WSN. Section 4 presents technical details of the smart contract including their deployment process and life cycle. Section 5 gives the implementation results and few tested scenarios of the approach. Section 6 concludes this work by indicating some drawbacks of this proposed model and some feature works.

2   Related Work

Securing WSN claimed to be one of the most interesting research because of their flexibility for problems solving in different domains obviously for the potential ones, that why it is widely used and involved in many industries such as IoT, ecommerce, pollution, science, marketing... etc. [11]. Because of the sensibility of the sent information from the WSN nodes [5], it is necessary to protect any piece of data using blockchain-based solution in order to produce better results and to prevent attacks such as selfish mining [4].

Blockchain-based solution is suggested in [9,10,12] by the authors for large-scale of IoT devices. Software Defined Network (SDN) is used for making decision and flexibility of attacks by constructing smart contract using ethereum technology in wide range in networks and blockchain. The provided model deals with the DDoS mitigation for searching the issues of the transfer of attack activities in decentralized networks. In [8], an computational resource sharing framework is investigated for D2D Network using blockchain [12], the authors described recuperation of failed nodes based on blockchain technology to recover nodes in terms of computational power, energy, and data storage in WSN looking for the state of cluster header by performing the security analysis. In [7], the authors presented a dynamic cluster head in distributed architecture of WSN approach in low energy cluster hierarchy (LEACH) aims to increase energy consumption during the selection of the CH operation. In [3], the authors proposed a new technique of cluster head selection in WSN using fuzzy decision-making by including energy consumption distance of nodes in the areas between neighbors nodes and base station in order to optimize the complexity of choosing cluster heads and decreasing the lifetime notwork.

To address the powerlessness of existing approaches, we proposed the cluster-Header selection in WSN based on blockchain technology using a smart contract to develop decentralized applications using ethereum environment. In fact, with all the mentioned features, we can explore all the included sections which collect for us much information including existed Blockchain accounts, written blocks, submitted transactions, deployed contracts, passed events, and logging information under logs. However, ensuring the stability, security against attacks, energy efficiency, and availability of cryptography mechanisms to avoid any modification of existing nodes due to tractability of each node in WSN.

3   Proposed Model

3.1   System Description

In this section, we describe the general architecture of our offered model. However, the idea is setting in place a new vision of security by taking all the advantages of Blockchain in order to prevent any kind of security issues. This architecture insures the non single control by any authority and improve the trustworthiness between parties by exploiting asymmetric cryptography system (private/public key encryption and decryption) to enhance the chances of any data immutability and arrange the way that data is transmitted between nodes over the network.

Along the way, blockchain is an innovative technology involved in many industries and domains with the aim of closing the road to any kind of data mutability of any piece of information that is already stored which makes this approach a great choice for our path [15]. Based on that our model is broken down to clusters and each cluster has various nodes, one of these nodes is a master one dedicated for receiving numerous data. The remaining nodes of a network (Cluster) will carry out of multiple tasks at once, on of these tasks is sending the same piece of information to the Master node and mine it on the blockchain directly. On the other hand, the master node will be charged of checking the validity of the received information by comparing them with the stored ones on the blockchain as shown in Fig. 1.

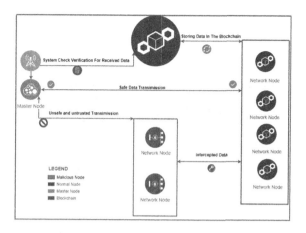

Fig. 1. Proposed model for securing cluster-header decision based on blockchain.

3.2 Formulation

We consider an area composed of one sink node, n nodes with m clusters, each cluster c is composed with x_c nodes and its cluster-header is denoted by CH_c as depicted in the Fig. 2. The energy of each node in a cluster is denoted by E_i^c and the minimal energy for the CH is denoted by E_{min}. Let $x_{i,j} = 1$ means that the node i belongs to cluster j else 0.

The first estimation brings the top node because at the beginning energies are equivalent. After that, decreasing the energy is related to the spent time and energy to behave and release demanded transactions over the network, defined as:

$$getEnergy(node.address) > max(E_1^c, E_2^c, ..., E_{x_c}^c). \tag{1}$$

Each node must exist in one cluster at most. So that, either a single node found in one place or not. This means that:

$$\sum_{j=1,i}^{m} x_{i,j} \leq 1, i = 1, ..., n. \tag{2}$$

Taking the energy of each node as an input and calculate the highest energy value of the network to decide which individual will be selected as a header for a given cluster to perform and done the execution of transactions, we verified:

$$E_{current} \geq E_{min}. \tag{3}$$

A CH is privileged based on its calculated energy which must exceed intensity conditions; the entire energies of all nodes situated in the network and the minimal one (3) as well to mine the next block in the Blockchain. As long as the network stills alive, the traffic will proceed it is way and the transactions ought to be validated and executed by the strongest node called Cluster-Header.

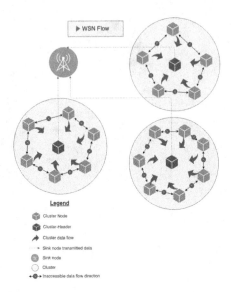

Fig. 2. An example of the WSN data flow.

However, freezing the transactions requires to set an end of the data flow over the chain. For this reason, we are assigning a variable named **transactionsEnd** which will be tested in the next appeal for a new transaction to perform it or not. If at least one node satisfies the energy condition, it will perform the next transactions. Otherwise, **transactionsEnd** will be set true. Therefore, the **Algorithm** 1 shows how the cluster head is selected based on their residual energy. Such as nodes with the largest residual energy amount are preferred to be cluster head.

4   Deployment of Smart Contract in WSN

A smart contract is a computer protocol expected to digitally facilitate, verify, or require the negotiation or performance of a contract. Smart contracts allow the performance of credible transactions without third parties; smart contracts render transactions traceable, transparent, and irreversible in WSN. The code and the agreements contained therein exist across a distributed, decentralized blockchain network.

In the beginning, one could consider a contract as a class in object-oriented terms, which is a template for objects. A contract may be deployed to a network multiple times, and each instance would have a distinct address, which could be used to interact with that particular instance of the contract at a later point. Therefore, each deployment of a contract could be considered as an object instance in oriented object concepts.

Algorithm 1 Dynamic change of Cluster-Header Decision Algorithm

---

Data: Nodes energy;
Result: Decide Cluster-Header
1 Initialization of the main parameters;
 Step1:
 Read minimal required energy (E_{min});
 Launch environment;
 Step2:
 while *all nodes energy are greater than the minimal value* **do**
2   │   **Step3:**
   │   Read current Cluster-Header energy ($E_{current}$);
   │   **if** $E_{current} > E_{min}$ **then**
3   │ │   **Step4:**
   │ │   Perform the transaction with the current node;
4   │   **else**
5   │ │   **Step5:**
   │ │   Set *isChanged = false* to check if the Cluster-Header is selected or not;
   │ │   **while** *Cluster-Header not yet selected* **do**
6   │ │ │   **for** *each node in the demanded cluster* **do**
7   │ │ │ │   **if** $getEnergy(node.address) > max(E_1^c, E_2^c, ..., E_{x_c}^c)$ **then**
8   │ │ │ │ │ set current loop node as the Cluster-Header; break loop;
9   │ │ │ │   **else**
10   │ │ │   **end**
11   │ │   **end**
12   │ │   **if** *isChanged == false* "Means no picked cluster-header" **then**
13   │ │ │ Set *transactionsEnd = true* "ie. freeze data flow over the cluster"
14   │ │   **end**
15   │   **end**
16 **end**

Furthermore, each instance is independent and has its state (persistent data). The constructor of the contract is invoked when deploying a contract to the network, and that is the only time it is invoked as depicted in Fig. 3.

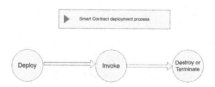

Fig. 3. Smart contract deployment process.

Creating a smart contract of WSN requires to know who is interacting with it at run time. In the ethereum blockchain, actors (smart contracts or wallets) are identified by their addresses. Storing addresses are used to handle the implementation logic depending on the transaction creator and its whole purpose as well. Destroy a contract is considered as a separated phase that can be done using a pre-built function mainly called **self destruct**. The smart contract life cycle is presented in Fig. 4.

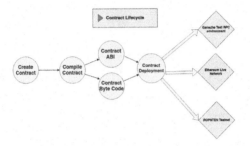

Fig. 4. Smart contract life cycle

Based on the desired business process, we will deploy our contract which contains a set of codes that satisfy our needs based on the application requirements. To follow the development process, we should retrieve the code of smart contract, then compile it and get the application binary interface which will be used to interact with our automate (smart contract). In Table 1, we are exploring our different built functions that perform different actions.

Table 1. Contract functions

Functions	Meaning
getSinkNode()	Retrieve sink node from the blockchain
sendData()	Send data from a node to an other in the same cluster
sendDataToSinkNode()	Send data from a cluster-header to the sink node
getTransactionsCount()	Get number of transactions stored in the contract for a given node
getSinkNodeTransactionsCount()	Get number of done transactions by the sink node stored in the contract
getGlobalTransactionsCount()	Get number of transactions stored in the contract
getClusterNodesAddresses()	Retrieve nodes of a given cluster
getTransactionsDate()	Retrieve transactions date of a given node
getAllTransactionsDate()	Retrieve all transactions date
getTransactionDetails()	Retrieve a transaction details
getClustersCounters()	Get number of clusters

In this study, we estimate the cost of the creation of the smart contract as well as the execution of each function used in WSN. We carried out the experiment values, when the gas price was set to $1Gwei$, where $1Gwei = 10^9 wei = 10^{-9}$, and $1ether$ was equal to $172,53USB$.

Table 2 illustrates the low cost of the running process of different functions in term of some indicators like the used Gas and the Fees used to construct

Table 2. WSN Creation and Low cost contract functions.

Function	Gas used	fee(ETH)	fee(USD)
constructor	6700000	0.02747	$6.23569
getSinkNode()	21943	0.00009	$0.02043
sendData()	334048	0.0013696	$0.3109
sendDataToSinkNode()	332819	0.0013646	$0.30976
getTransactionsCount()	50213	0.0002059	$0.04674
getSinkNodeTransactionsCount()	57927	0.0002375	$0.05391
getGlobalTransactionsCount()	27421	0.0001124	$0.02551
getClusterNodesAddresses()	207357	0.0008502	$0.193
getTransactionsDate()	84317	0.0003457	$0.07847
getAllTransactionsDate()	43146	0.0001769	$0.04016
getTransactionDetails()	31384	0.0001287	$0.02921
getClustersCounters()	21902	0.0000898	$0.02038

our contracts. We observe that the highest-paid amount is corresponding to the deployment function only with 6.23569 USD. Nonetheless, it is only executed while setting up the system collaboration. As well as, remaining functions use approximately an amount between 0.02 and 0.3 USD as a maximum paid amount.

5   Implementation

5.1   Tools

Before we start, a set of dependencies are needed for development reasons. The contract must be consistent with the business process of the web application to achieve satisfactory results. First of all, we need to install Python as an interpreter to deploy the smart contract developed by the programming language solidity [15]. As presented above, we described the parameters used to provide clarity on the work of the ecosystem:

- **Smart contract:** an Ethereum smart contract is bytecode deployed on the Ethereum blockchain. There could be several functions in a contract.
- **Django:** Django is a Python-based free and open-source web framework, which follows the model-template-view architectural pattern. It is maintained by the Django Software Foundation, an independent organization established as a 501 non-profit. Django's primary goal is to ease the creation of complex, database-driven websites.
- **ABI:** an ABI (Application Binary Interface) is necessary to specify which function in the contract to invoke, as well as get a guarantee that the function will return data in the format we are expecting.

- **Web3:** web3 is a collection of libraries that allow us to interact with a local or remote Ethereum node, using an HTTP, WebSocket or IPC connection.
- **Solc:** solidity compiler is the special program that processes statements written in solidity programming language and turns them into machine language or "code"' that a computer's processor uses.

The parameters values used in our contribution are recorded in Table 3.

Table 3. Cluster Head selection Parameters.

Parameters	Values
Total number of nodes	270
Size of the monitoring area	200 m * 200 m
Number of clusters	6
Balance	100 (Initial Value)
Block hash	Ethash
Gas used	26868
Gas price	20000000000
Gas limit of blocks	6721975
Gas limit of transactions	90000

5.2   Balancing Energy

In this part, several test of scenarios are investigated to ensure the proper functioning of the proposed model.

We consider the transactions are not directly stored in the contract but they construct a supply-chain called blockchain. More precisely, the transaction goes through a series of steps before being completed. First of all, we need to check the current Cluster-Header (CH) by verifying its energy if it satisfies the minimum required one as specified in the Algorithm 1. Knowing in advance the full network nodes has the advantage to proceed on a fully recognized chain by injecting necessary data in the contract itself without need to scan the network before each transaction. In the mean time, less time is consumed during any transaction thanks to the non complexity activities. Not only that, but also the number of cluster is determined from the beginning.

Typically, when we access a multi-nodes system, we access either a master node or a gateway node (default one). The CH is configured to be the start point for the jobs running on the network. When a user desires to log in or access to the system, he is automatically prompted to log on to the primary node. In the blockchain, a transaction is performed by the CH which is changed when its energy reaches a minimum balance leading to the end of the data transmission when all the nodes have a balance lower than the minimal one.

Figure 5 shows several running of test scenarios such as the selection of cluster-header node, the sink node, and the transaction between nodes.

Fig. 5. Successful data transmission scenarios.

The simulation was performed to ensure the dynamically functioning of the platform to execute a full scenario. As soon as the user (node) is connected with his blockchain address, he will be able to perform a transaction in the deployment network by sending any kind of data. In the main context, sending data always takes the path to the CH which is chosen dynamically, then the CH itself will perform the remaining transaction to the sink node which receives data only from CHs of different clusters. Transactions mining as shown in Fig. 6 is the main operation for the whole process. In this context, when transferring data over the network we are manipulating security aspects of cryptography system such as private/public encryption with the aim of storing data respecting the written rules on the contract and storing data inside the network for later check form all parties. This action has the goal of preventing any kind of modification or fraud thanks to the immutability principal.

The main goal consists of performing any transaction by the powerful and robust node of the network. This node in reality is chosen based on its energy level which must exceed the whole nodes one which are placed in the same cluster. The methodology of our model is very flexible on the variety of taking a node as a CH one. The reason of taking the powerful node from the network is based on the insurance and quality criteria, that's why in cluster1, cluster2, and cluster3 is shown in Table 4, we are taking the node with the largest amount of the residual energy (as shown in the last account address will be selected) to be responsible for transferring data from the cluster to its target.

Checking these criteria is allowing to take the advantages of system stability over time from any future changes, security between nodes, and availability even in critical situations.

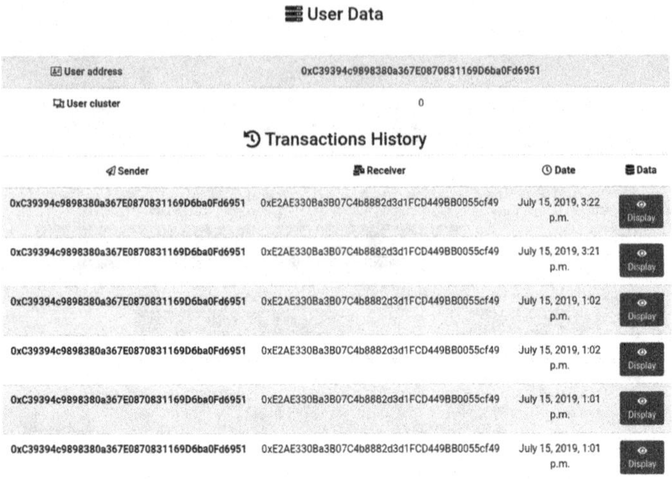

User Data

| User address | 0xC39394c9898380a367E0870831169D6ba0Fd6951 |
| User cluster | 0 |

Transactions History

Sender	Receiver	Date	Data
0xC39394c9898380a367E0870831169D6ba0Fd6951	0xE2AE330Ba3B07C4b8882d3d1FCD449BB0055cf49	July 15, 2019, 3:22 p.m.	Display
0xC39394c9898380a367E0870831169D6ba0Fd6951	0xE2AE330Ba3B07C4b8882d3d1FCD449BB0055cf49	July 15, 2019, 3:21 p.m.	Display
0xC39394c9898380a367E0870831169D6ba0Fd6951	0xE2AE330Ba3B07C4b8882d3d1FCD449BB0055cf49	July 15, 2019, 1:02 p.m.	Display
0xC39394c9898380a367E0870831169D6ba0Fd6951	0xE2AE330Ba3B07C4b8882d3d1FCD449BB0055cf49	July 15, 2019, 1:02 p.m.	Display
0xC39394c9898380a367E0870831169D6ba0Fd6951	0xE2AE330Ba3B07C4b8882d3d1FCD449BB0055cf49	July 15, 2019, 1:01 p.m.	Display
0xC39394c9898380a367E0870831169D6ba0Fd6951	0xE2AE330Ba3B07C4b8882d3d1FCD449BB0055cf49	July 15, 2019, 1:01 p.m.	Display

Fig. 6. Transactions history of accounts nodes

Table 4. Cluster-header node flexibility in term of energy.

Account address	Energy	TX Count	Index
Cluster1			
0x8841aA78426f1077CF7E2789E5f7748C4463C19F	96.90	76	0
0xE2AE330Ba3B07C4b8882d3d1FCD449BB0055cf49	99.48	68	1
0xC39394c9898380a367E0870831169D6ba0Fd6951	99.86	18	2
0x3474fA62123D4497257223e9169De2AA98862Ae9	99.98	3	3
0xA3747842C6e5fC5903C12b7098d87DfEc3bc293f	99.94	13	4
0x5288a3ccE3F45be0C374072dccAD829B9f8DFe91	99.99	1	5
Cluster2			
0x180524eD52a43b5d3e1E88dae42d486c122bd3BB	99.93	97	0
0x1C10dE035539a5cdb39C5BEcaF767f78adeF6844	99.94	94	1
0x0304E8A643bd5a8401dc0250db472a00706d5939	99.95	105	2
0x00426aE247e572Ab18318219E6c09a300F4bF31B	99.96	96	3
0xBE14431B101bE86987283834929ceFA03b862c16	99.98	121	4
0x4bd3CD24c12F9dc197582d82D87c4439cfC8733d	100	124	5
Cluster3			
0xf5cAF46A61Ef892Cabb7804bd3AC42b3e30e80E3	97.80	17	0
0xB78e22f1896bcA33c1003E8eCd73a958Ae85af50	98.95	43	1
0x802A15465a54f854O66b3D301503d6A0C0e4ACD4	99.90	20	2
0x182f7Cc7dAa77bc036cfB22a59D84089f0fG51D3	99.94	6	3
0x25327a33Ec5ce2067921085436Bc0a45539Oc9f8f	99.95	34	4
0x6B0a718F9c72b19c8Ec89B858c4C0a283e6F54cd	99.96	50	5

Figure 7 shows the energy-consumption state of the entire wireless sensor network node. The comparative results of the energy consumption status, the LEACH was used 0.0347J, the DDACM was used 0.0265J, the RSSI cluster routing method was used 0.0248J, and our proposed scheme was 0.0237J. Thus, it were could be obtained result of energy consumption reduced compared with existing approaches of cluster head selection in WSN network.

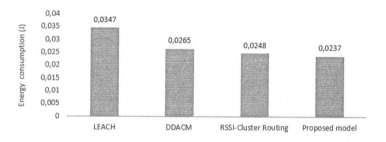

Fig. 7. Energy-consumption state of nodes for each approach.

6 Conclusion

In this study, we have proposed a secured framework using the ethereum (ETH) smart contract tool to choose the Cluster-Header (CH) that can be integrated with the previous model/existing to designate a CH from a set based-blockchain clusters to guarantee the security of data transmission from sensing nodes to the sink through cluster heads with low energy consumption comparing with the existing approaches and low-cost for deploying the smart contract is performed. Therefore, ensuring the system availability and the safety of database decentralization. Whereas, we may have some drawbacks in our suggested framework like performing transactions with massive data (images, videos, texts) is surely costing high price procedure.

As future work, we will set the target of the integration of our framework with others' process of safety and applied it in decentralized architectures in another network than WSN, to empower and ensure the security and the privacy of the whole framework. In addition, we will set the necessary performance indicator of this approach based on the comparison with other cryptocurrencies like Bitcoin, Litecoin.

References

1. Abdullah, K.M., Houssein, E.H., Zayed, H.H.: New security protocol using hybrid cryptography algorithm for WSN. In: 1st International Conference on Computer Applications & Information Security (ICCAIS), pp. 1–6. IEEE (2018)

2. Al-Karaki, J.N., Gawanmeh, A., Ayache, M., Mashaleh, A.: Dass-care: a decentralized, accessible, scalable, and secure healthcare framework using blockchain. In: 15th International Wireless Communications & Mobile Computing Conference (IWCMC), pp. 330–335. IEEE (2019)

3. Azad, P., Sharma, V.: Cluster head selection in wireless sensor networks underfuzzy environment. ISRN Sensor Netw. **2013**, 909086 (2013)

4. Bai, Q., Zhou, X., Wang, X., Xu, Y., Wang, X., Kong, Q.: A deep dive into blockchain selfish mining. In: IEEE International Conference on Communications (ICC), ICC 2019, pp. 1–6. IEEE (2019)

5. Bloch, M., Barros, J., Rodrigues, M.R.D., McLaughlin, S.W.: Wireless information-theoretic security. IEEE Trans. Inf. Theory **54**(6), 2515–2534 (2008)

6. Cheikhrouhou, O., Koubaa, A.: BlockLoc: secure localization in the internet-of-things using blockchain. CoRR abs/1904.13138 (2019)

7. Gupta, K., Goyal, A., Tripathi, A.K.: A novel approach for cluster head selection in wireless sensor network. Int. J. Comput. Appl. **113**(16), 22–27 (2015)

8. Hong, Z., Wang, Z., Cai, W., Leung, V.: Blockchain-empowered fair computational resource sharing system in the D2D network. Future Internet **9**(4), 85 (2017)

9. Makhdoom, I., Abolhasan, M., Abbas, H., Ni, W.: Blockchain's adoption in IoT: the challenges, and a way forward. J. Netw. Comput. Appl. **125**, 251–279 (2019)

10. Misic, J., Misic, V.B., Chang, X., Motlagh, S.G., Ali, M.Z.: Block delivery time in bitcoin distribution network. In: IEEE International Conference on Communications (ICC), ICC 2019, pp. 1–7 (2019)

11. Nakamoto, S.: Bitcoin: a peer-to-peer electronic cash system. Technical report, Manubot (2019)

12. ul Hussen Khan, R.J., Noshad, Z., Javaid, A., Zahid, M., Ali, I., Javaid, N.: Node Recovery in Wireless Sensor Networks via Blockchain. In: Barolli, L., Hellinckx, P., Natwichai, J. (eds.) 3PGCIC 2019. LNNS, vol. 96, pp. 94–105. Springer, Cham (2020). https://doi.org/10.1007/978-3-030-33509-0_9

13. Riabi, I., Ayed, H.K.B., Saidane, L.A.: A survey on blockchain based access control for internet of things. In: 15th International Wireless Communications & Mobile Computing Conference (IWCMC), pp. 502–507. IEEE (2019)

14. Sayadi, S., Rejeb, S.B., Choukair, Z.: Anomaly detection model over blockchain electronic transactions. In: 15th International Wireless Communications & Mobile Computing Conference (IWCMC), pp. 895–900. IEEE (2019)

15. Spathoulas, G., et al.: Towards reliable integrity in blacklisting: facing malicious IPs in GHOST smart contracts. In: Innovations in Intelligent Systems and Applications (INISTA), pp. 1–8. IEEE (2018)

Security in MANETs: The Blockchain Issue

Nada Mouchfiq$^{(\boxtimes)}$, Chaimae Benjbara , and Ahmed Habbani

ENSIAS, Mohammed V University in Rabat, Rabat, Morocco
mouchfiq.nada@gmail.com, c.benjbara@um5s.net.ma, ahmed.habbani@um5.ac.ma

Abstract. With the emergence of the IoT (Internet of Things) technology, security features must take into account the developments and possible uses. The blockchain is considered as a new technology devoted to data sharing, it allows to guarantee the integrity of data by IoT devices thanks to intelligent contracts allowing to verify and enforce the conditions of the monitored goods during their processing. In our work, we will address the security of Blockchain-based systems and networks to improve and simplify the communication of messages and informations, using the most innovative and sophisticated processes and a form of device independent control. In order to do so, we will analyse the solutions advanced by researchers from all over the world who are dedicated to guaranteeing a good state of security on these networks. Finally, our manuscript will present an approach to enhance security which is founded on the Blockchain principle and which we have entitled "MPR Blockchain". This solution proves to be the most appropriate to our requirements and expectations as a team working within IoT systems and especially ad hoc networks.

Keywords: Ad hoc networks · Blockchain · MPR · Security

1 Introduction

It has become essential to show all members of the community on the need to preserve safe and personal sides; as long as there is a sensible environment, [3] advanced and improved information technologies and communications which can save time and deliver results faster. This leads us to remember the safety of the environment Digital intelligence and privacy within the planet, there is a change of confidence within the reliability of the online.

The current research is oriented towards the phenomenon of the Internet of Things (IoT) associating both hard and soft aspects [12]. The mixture of the web and emerging technologies has identified the concept of digital environment.

The Internet of Things (IoT) is a concept that establishes connections between several components within the Internet network and gathers all the data exchanged by different smart appliances using integrated systems, sensors, software and AI systems and isn't surrounded on some devices or technologies but this concerns several objects of various types.

© Springer Nature Switzerland AG 2020
M. Belkasmi et al. (Eds.): ACOSIS 2019, CCIS 1264, pp. 219–232, 2020.
https://doi.org/10.1007/978-3-030-61143-9_18

Nowadays, the phenomenon of data and technology infrastructures and vast flows of data became essential in defining modern life. Each challenge represents an ample and discreet notion where know-how and leadership in welfare work are often focused on new bold ideas, scientific discoveries, and surprising innovations [13]. Within the coming years, IoT will link several intelligent devices.

As a researchers, we work on improving the performance of ad hoc networks (MANETs, VANETs, FANETs, ...). Since ad hoc networks could integrate the new IoT technology, we'll be ready to cash in of this sort of network to enhance ours. During this article, we'll study security improved by the principle of blockchain to seek out a relevant solution increasing the safety level of ad hoc networks.

The security issues of the Internet of Things can be cited as follows:

- Physical security that focuses on sensor protection and nodes in the case of interaction, signal presence including interference and sensor communication.
- Security that is linked to the various sensor functions and is very similar to the safety process linked to the assorted sensor tasks, which are inherently similar to security in conventional communication systems.
- Data privacy extends to a range of components, such as the security of communications in the sensor nodes, the broadcasting channels and the reputation system.

To make the network more flexible and secure, blockchain is a range of methods used within distributed systems [18] in order to establish a consistent and secure basis of data between all the components of the network.

This paper is organized as follows. Section 2 gives an overview about IoT architecture, counter-measures and security. Section 3 discusses Blockchain principle and security T. In Sect. 4, we discuss the related works which lead us to our proposed solution in Sect. 5. Finally, our research is concluded in Sect. 6.

2   IoT

2.1   IoT Definition

Internet of Things (IoT) is taken into account as emerging concepts and technologies [8]. At an equivalent time he tries to rework certain concepts from which he can create new possibilities, with adapted scenarios.

The Internet of Things is defined as a big entity that has services and technologies that allow communication between its various components [11], the IoT contributes to the optimization or the creation of latest concepts whether for the businesses or for people and in several areas namely: mobility, transport, smart city, distraction, architecture, industry, health ..., etc.

2.2   Challenges of IoT

The growing prevalence of intelligent systems embedded in virtually all kinds of consumer devices and therefore the criticality of certain applications (such as monitoring, online health and network control) dictate the necessity for reliable security [15]. The issues related to reliable security in IoT are motivated by the subsequent factors:

- Problem of understanding: IoT systems are complex and less understandable than normal systems.
- Distribution problem: IoT systems are known by their distribution over huge areas and in most cases are not sufficiently secure or controlled.
- Addressing Problem: IoT systems opt for various endpoints and use multiple addressing schemes as needed, which makes these systems complex.
- Problem related to the limitation of battery: This is due to the variety of the functionalities of the IoT systems and which have a rather limited battery.
- Architecture issue: IoT systems do not have standardized security tools and techniques and so far this principle has not been implemented yet.

2.3 IoT Architecture

The architecture of the IoT network consists of 4 layers as shown in Fig. 1. This is often not a typical architecture for IoT [16] but we'll adopt it as a reference architecture since it's the foremost used and most accepted so as to spot and classify different security issues in IoT [1].

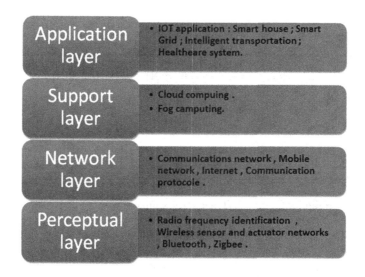

Fig. 1. IoT architecture

2.4 IoT Countermeasure

The countermeasure is that the assembly of actions highlighted in mistrust to stop certain risks.

The countermeasures [23] to be highlighted aren't only technical resolutions, but also training and awareness-raising measures for users, also as a clearly defined set of rules.

In order to be ready to make a system reliable, it's necessary to differentiate the potential threats, and thus to assimilate and to experiment the way of execution of the enemy. The target of this file is thus to offer an estimate of possible hackers, to categorize them, and eventually to offer an summary of the way to reduce the risks of intrusions as shown in (Fig. 2) below.

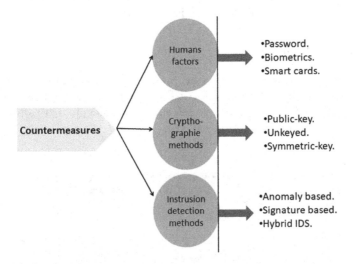

Fig. 2. IoT countermeasures

2.5 IoT Security Criteria

The key requirements for any IoT security solution are:

- Authenticity: The only people who could have authorized access to the system are the legal users.
- Authorization: permissions and rights granted components so that they only have access to the resources to which they are licensed.
- Confidentiality: The nodes through which information is transmitted must be protected against intrusion.
- Integrity: Associated data must not be altered.
- Availability and continuity: These two criteria are essential in order to avoid any lack of fulfillment and interruptions as exploitable as potential.

2.6   IoT Security Technics

IoT network security is more complex than traditional network security because the IoT network has more communication protocols, additional standards and features, and this complexity results in hackers and malicious devices which will harm the network [17]. The robust security techniques for IoT networks include:

- IoT authentication: Allows the management of several devices in the system either through passwords or simple keys or biometrics. In the context of the IoT, the authentication scenarios do not require the Human intervention, only machines can take care of it.
- IoT Encryption: Allows to reserve the integrity of the information in the network and makes it difficult to hack the system, that is, there is no standard encryption process in the IoT but what is common is the process life cycle and is essential for security management in general.
- IoT PKI: In the same framework of the concept of life cycle, the generation of private/public keys provided by the PKI since the hardware rating can restrict the choice of the key and these allow to ensure the desired level of security.
- IoT security analysis: This is an approach that aims to make operations at the level of IoT devices, namely: the collection, aggregation, monitoring and standardization of data and the creation of reports and alerts. This IoT security analysis approach detects attacks and intrusions that are not affected by traditional network security solutions.
- IoT API security: This notion is essential to ensure the integrity of the information flowing through the network and, to detect attacks that can threaten the system and to authenticate and authorize data matching between devices.

3   Blockchain

3.1   Blockchain Definition

The blockchain is a technology for storing information and transmitting in an exceedingly transparent, secure and decentralized way. It is a kind of large database that stores the archives of all the transactions conducted among its users since the creation of the blockchain. The good feature of the blockchain is its decentralized architecture as we mentioned above, actually it's hosted by one server but by some users. The components of the blockchain don't need intermediaries in order that they will verify the validity of the chain and therefore the information and are equipped with security procedures that protect the system [20].

Transmissions between network users are grouped in blocks and every of those blocks is validated by network nodes called minors, supported criteria that depend upon the sort of blockchain. Once the block is validated, it joins the opposite blocks and is included in the blockchain. The operation is then shown to the destination, as well as to the whole system (Fig. 3 and 4).

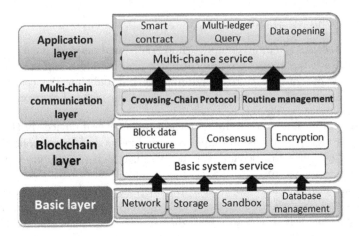

Fig. 3. Blockchain architecture

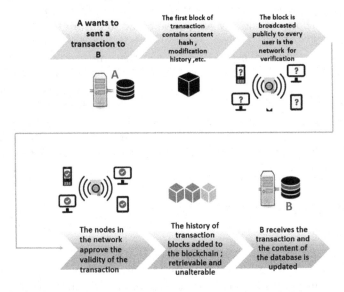

Fig. 4. Operating principle of the blockchain

Since it is a technology that stores and protects data on a large scale, the blockchain is used in several fields including:

- Online trading
- Internet of things
- Cybersecurity
- Bank transfers and management
- Voting systems
- Smart agriculture
- Smart Grid

– Health
– Management of large groceries

3.2 Blockchain's Security

A blockchain is a chain of blocks which contain specific information (database). In the case of the distributed chain architecture network, every member in the network retains, validates and upgrades the updated inputs.

The blockchain takes the format of a variety of methods applied in distributed networks in order to ensure that a coherent database is held by all members. It is primarily advanced by Satoshi Nakamoto [21] to encapsulate the fundamental concepts of the established digital currency techniques, that is to say the Bitcoin. In opposition to the traditional centralized network structure, there are no fixed central nodes in networks based on blockchain.

All the members of the network have reasonably equitable roles and stock an appropriate duplicate of blockchain. Thanks to the increased level of security and trustworthiness, the blockchain has been used in several applications scenarios and is taken into account one among the key techniques to promote the development of the world.

Blockchain mechanisms are ideal for this requirement as they support authentication and authorization, but availability are often provided by intrusion detection systems. Authorization and authentication could also be considered an integral part of the integrity requirement.

When multiple nodes have an identical block in their principal chain, they are treated as having achieved a consensus. This sub-section outlines the approval requirements for every block and how consensus is obtained and sustained. We also outline other consensus methods that are presently in practice.

The consensus process includes two phases: the approval of the blocks and consequently the election of the most expanded chain, these two phases are executed autonomously by every node. The blocks are spread over the network, and then every node which gets a substitute block retransmits it to its neighbours. But prior to this redistribution, the node executes a block check to guarantee that only validated blocks are distributed. There is an exhaustive checklist to be respected, namely:

– Checking whether the header hash satisfies the specified complexity
– Block dimensions in the expected boundaries
– Block structure
– Checking the timestamp
– Verification of all transactions

4 Related Work

4.1 Towards Better Availability and Accountability for IoT Updates by Means of a Blockchain

This approach [6] highlights the usefulness of the blockchain in the case of downloads of updates especially in a large IoT network since in the case of an upgrade included in the blockchain using an approved block, the attacker can not delete the block. The proposed method takes into account the unchanged characteristics of the blockchain which is then applied to check that these changes and upgrades are protected after a consensus procedure among IoT appliances and thus ensures the stability of the network during the update and increases the life of the devices by avoiding any unnecessary waste of resources.

4.2 Blockchain-Based Asymmetric Group Key Agreement Protocol for Mobile Ad Hoc Network

The authors [22] proposed a solution based on anonymous authentication technology in the advancement of privacy protection in the authentication process of every participant. The solution proposed by the authors is generally used in the limited mobile network, and also use the techniques available in blockchain technology to ensure traceability and improve liability that can not be made without being detected. This solves the problem of technical strangulation in an ad hoc network, and satisfies the key information exchange application in this network environment.

4.3 Supporting Connectivity of VANET/MANET Network Nodes and Elastic Software-Configurable Security Services Using Blockchain with Floating Genesis Block

The authors of this work [7] proposed to use Blockchain technology adapted to the topology of the VANET/MANET network, the distribution and storage of authentication data in this type of network. The issue of the unrestrained increase in the blockchain that has inhibited the deployment of this approach in the VANET/MANET networks is investigated. The process was to propose a blockchain modification based on a floating genesisblock to solve growth problems with unlimited blockchain and with secure routing protocols on Blockchain technology. This securing allows the connection of VANET/MANET network nodes with software defined security services as well as system control in general.

4.4 Towards an Optimized BlockChain for IoT

This approach [9] consists of integrating a lightweight blockchain architecture dedicated to IoT by eliminating the overhead expenses of the current blockchain while providing the security and the confidentiality as a solution, since the blockchain principle necessitates a high calculating capacity and a high calculating time. To do this, the author proposed a blockchain architecture with a centralized node to optimize the efficiency of the batteries and distributed approvals to reduce processing time and block checking, in addition to the implementation of functional blocks guaranteeing security and confidentiality.

4.5   Secure IoT Communication Using Blockchain Technology

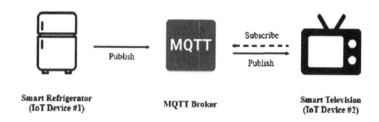

Fig. 5. IoT system without blockchain technology

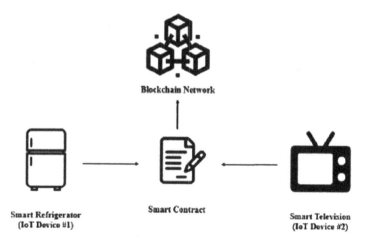

Fig. 6. IoT system using blockchain technology

This proposal [10] resolves security issues that affect communication between the various components of the IoT network since blockchain integration makes the network security level higher than that of an IoT network without blockchain.

Indeed, after analyzing these two IoT systems (without blockchain Fig. 5 and with blockchain Fig. 6), in order to determine the security level of each, the author simulated attacks and observed security features. The testing findings show that in fact, the IoT system that uses blockchain has a significantly greater degree of security compared to the IoT system that does not use blockchain.

5   Proposed Solution

5.1   Discussion

Nowadays, the phenomenon of data and technology infrastructures and vast flows of data became essential in defining of recent life. Each challenge represents an ample and discreet notion where know-how and leadership in welfare work are often focused on new bold ideas, scientific discoveries and surprising innovations [19]. Within the coming years, IoT will make the link between several intelligent devices.

The Internet of things refers to a type of network to connect anything with the Internet. This environment is composed of differents networks (Cloud, Big Data, Industry 4.0,..) among them we find ad hoc networks.

Several approaches have been suggested for merging mobile ad hoc networks with the Internet. As the nodes of mobile ad hoc networks have IP addresses to define their routing, it might be appropriate to consider the use of these IPs to route a packet over the Internet.

An ad hoc network is a kind of decentralised wireless network that does not rely on any pre-existing infrastructure, be it routers or access points. In the ad hoc networks each node contributes in the routing by retransmitting the data to the other nodes, and the choice of the node that will transmit the data is operated in a dynamic manner using the network connectivity in accordance with the routing algorithm applied.

There are several types ad hoc networks: MANETs (Mobile ad-hoc networks), VANETs (Vehicular ad-hoc networks), FANETs (Flying ad-hoc networks), BANETs (Body ad-hoc networks). Each network has his own challenges, ad hoc networks can face several challenges including:

- Energy constraint.
- Constraint of absence of fixed limits.
- Lack of protection against external signals.
- Routing constraint with the dynamics of the network topology.
- Constraint related to variation of link and device capabilities.

The mobile ad-hoc networks (MANETs) are a system of mobile nodes connected with each other through wireless links without infrastructure maintenance. They are a tangible example of smart environments, as a set of wireless mobile nodes capable of communicating without pre-existing infrastructure. Each node forming this type of network contributes to the routing procedure.

As a researchers team, we focus on ad hoc networks, particularly on mobile ad hoc networks (MANETs), which are wireless networks with mobile elements without infrastructure and deliver the configuration on a continuous foundation. Each MANETs component has the ability to manage easily within the network and must whenever have traffic associated with its application and this is often what gives the MANETs the particularity of operating both as a router and as a host. We work on Optimized Link State Routing (OLSR) protocol, which is a reputable MANETs protocol and which approved particular nodes called

Multipoint Relays (MPRs) whose function is to limit the flow of broadcasting packets in the network by restricting repetitive rebroadcasts to the same node (Fig. 7).

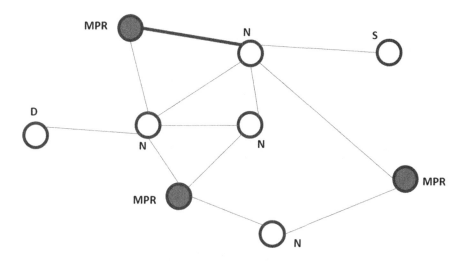

Fig. 7. Traditional MANETs network

These particular nodes are accountable for the optimization of TC messages. The interaction between the source (S) and therefore the destination (D) is afforded via intermediate nodes (N) which pick relay points among them established several properties like energy state, the stability of the node, and view of developing control messages along the network in an optimized way.

We are working on enhancing the OLSR protocol in MANETs networks by introducing solutions to handle the challenges of this kind of network. Among the work accomplished, some study the optimization of the MPRs selection based on various criteria for example mobility quantification [4] and decreasing broadcast redundancy [5], improving communication [2] and security [14].

According to our research, it's been found that there are new security methods implemented in other networks like the blockchain in IoT we've mentioned in our article. It's therefore been proposed to use the principle of blockchain security in ad hoc networks to determine its influence on the level of security that it'll offer to our MANETs network.

To ensure the continuity of the teamwork and especially to guarantee security for the network using the improved OLSR protocols elaborated in our team, we introduce a method of security more suitable to our requirements, this approach is founded on the Blockchain principle.

5.2   Our Proposition

In our "MPR Blockchain" approach integrating blockchain in a MANETs network, the MPR takes the minor position. After the MPR is nominated by the nodes (N) forming its neighbourhood and all of which are provided with utilizing a blockchain security algorithm. So values are assigned to them namely (Fig. 8):

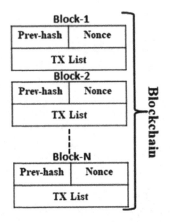

Fig. 8. Blockchain components

- TX list: transaction history of transactions disseminated on the network.
- Nonce: occurs in every block as an answer to the mathematical question.
- Previous hash: value relating to the precedent node.

As presented within the figure above (Fig. 9), after storing the values associated to each of the (N)in the MPR, the choice of the hash algorithm occurs

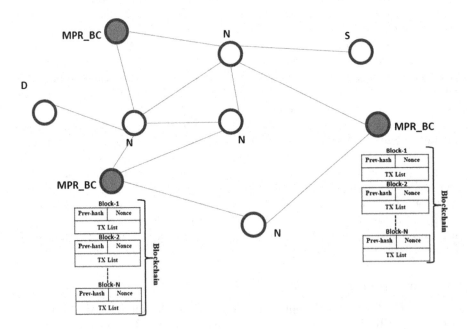

Fig. 9. "MPR Blockchain" integrated in a MANETs network

for generating the keys to build up and finish the entire package structure of our blockchain. Every MPR substitutes its own blockchain with the surrounding MPRs within the network and of course all of them are supporting transaction rules of the blockchain to make sure security within the network is guaranteed, we will name them "MPR Blockchain". And as a final step, every "MPR Blockchain" calculates the reliability estimated value of the network part which makes it willing to recognize malicious nodes within his neighborhood.

6 Conclusion

In this article, we presented the description of IoT, its architecture and infrastructure, and therefore the security of IoT by mentioning security measures. We also revealed the work already wiped out the world of network security applying blockchain technology then introduced our proposal. Our following work will be about treating the implementation and simulation part of the "MPR Blockchain" solution within the MANETs and analyze the accessed results with the recent achievements provided by the different members of the team operating on the section dealing with security.

References

1. Ali, I., Sabir, S., Ullah, Z.: Internet of things security, device authentication and access control: a review. arXiv preprint arXiv:1901.07309 (2019)
2. Benjbara, C., Habbani, A.: Communication in a heterogeneous ad hoc network. Int. J. Comput. Inf. Eng. **13**(10), 524–529 (2019)
3. Benjbara, C., Habbani, A., Mahdi, F.E., Essaid, B.: Multi-path routing protocol in the smart digital environment. In: Proceedings of the 2017 International Conference on Smart Digital Environment, pp. 14–18 (2017)
4. Berradi, H., Habbani, A., Benjbara, C., Mouchfiq, N., Souidi, M.: Enhanced mobile network stability using average spatial dependency. In: Ben Ahmed, M., Boudhir, A.A., Santos, D., El Aroussi, M., Karas, İ.R. (eds.) SCA 2019. LNITI, pp. 495–509. Springer, Cham (2020). https://doi.org/10.1007/978-3-030-37629-1_36
5. Berradi, H., Habbani, A., Benjbara, C., Mouchfiq, N., Souidi, M.: Enhancement wireless network stability based on spatial dependency. In: Proceedings of the 4th International Conference on Smart City Applications, pp. 1–6 (2019)
6. Boudguiga, A., et al.: Towards better availability and accountability for IoT updates by means of a blockchain. In: IEEE European Symposium on Security and Privacy Workshops (EuroS&PW), pp. 50–58. IEEE (2017)
7. Busygin, A., Kalinin, M., Konoplev, A.: Supporting connectivity of vanet/manet network nodes and elastic software-configurable security services using blockchain with floating genesis block. In: SHS Web of Conferences, vol. 44, p. 00020. EDP Sciences (2018)
8. Chourabi, H., et al.: Understanding smart cities: an integrative framework. In: 45th Hawaii International Conference on System Sciences, pp. 2289–2297. IEEE (2012)
9. Dorri, A., Kanhere, S.S., Jurdak, R.: Towards an optimized blockchain for IoT. In: IEEE/ACM Second International Conference on Internet-of-Things Design and Implementation (IoTDI), pp. 173–178. IEEE (2017)

10. Fakhri, D., Mutijarsa, K.: Secure IoT communication using blockchain technology. In: 2018 International Symposium on Electronics and Smart Devices (ISESD), pp. 1–6. IEEE (2018)
11. Forsyth, T.: Community-based adaptation: a review of past and future challenges. Wiley Interdisc. Rev. Climate Change **4**(5), 439–446 (2013)
12. Li, S., Da Xu, L., Zhao, S.: 5G internet of things: a survey. J. Ind. Inf. Integr. **10**, 1–9 (2018)
13. Liu, Y., Kuang, Y., Xiao, Y., Xu, G.: SDN-based data transfer security for internet of things. IEEE Internet Things J. **5**(1), 257 (2018)
14. Mahdi, F.E., Habbani, A., Mouchfiq, N., Essaid, B.: Study of security in MANETs and evaluation of network performance using ETX metric. In: Proceedings of the 2017 International Conference on Smart Digital Environment, pp. 220–228 (2017)
15. Minoli, D., Occhiogrosso, B.: Blockchain mechanisms for IoT security. Internet Things **1**, 1–13 (2018)
16. Nguyen-Minh, H.: Contribution to the intelligent transportation system: security of safety applications in vehicle ad hoc networks. Avignon (2016)
17. Ribeiro, S.L., Nakamura, E.T.: Pseudonymization approach in a health IoT system to strengthen security and privacy results from OCARIoT project. In: Doss, R., Piramuthu, S., Zhou, W. (eds.) FNSS 2019. CCIS, vol. 1113, pp. 134–146. Springer, Cham (2019). https://doi.org/10.1007/978-3-030-34353-8_10
18. Swan, M.: Blockchain thinking: the brain as a decentralized autonomous corporation [commentary]. IEEE Tech. Soc. Mag. **34**(4), 41–52 (2015)
19. Vukobratovic, D., et al.: Condense: a reconfigurable knowledge acquisition architecture for future 5G IoT. IEEE Access **4**, 3360–3378 (2016)
20. Wessling, F., Gruhn, V.: Engineering software architectures of blockchain-oriented applications. In: IEEE International Conference on Software Architecture Companion (ICSA-C), pp. 45–46. IEEE (2018)
21. Yang, X., Lau, W.F., Ye, Q., Au, M.H., Liu, J.K., Cheng, J.: Practical escrow protocol for bitcoin. IEEE Trans. Inf. Forensics Secur. **15**, 3023–3034 (2020)
22. Zhang, Q., Li, Y., Li, J., Gan, Y., Zhang, Y., Hu, J.: Blockchain-based asymmetric group key agreement protocol for mobile ad hoc network. In: Meng, W., Furnell, S. (eds.) SocialSec 2019. CCIS, vol. 1095, pp. 47–56. Springer, Singapore (2019). https://doi.org/10.1007/978-981-15-0758-8_4
23. Zhang, Z.K., Cho, M.C.Y., Shieh, S.: Emerging security threats and countermeasures in IoT. In: Proceedings of the 10th ACM Symposium on Information, Computer and Communications Security, ASIA CCS 2015, pp. 1–6. Association for Computing Machinery, New York (2015). https://doi.org/10.1145/2714576.2737091

Improving IoT Security with Software Defined Networking (SDN)

Abdelaali Tioutiou$^{(\boxtimes)}$ and Ouafaa Diouri

Ecole Mohammedia Des Ingénieurs, Mohammed V University in Rabat,
Rabat, Morocco
abdelaalitioutiou@research.emi.ac.ma,
diouri@emi.ac.ma

Abstract. The presence of Internet of Things (IoT) in our daily lives becomes a reality and its development has become truly remarkable, indeed the massive use of connected devices has disclosed the importance of the security issues related to the processing of data collected and/or provided by these devices. In this sense, the dynamic configuration of network becomes a necessity. Software Defined Networking (SDN) is a new paradigm, which allows the centralized management of network devices and it provides new mechanisms to solve some problems related to networking.

In this paper, we focus on the benefits provided by SDN paradigm to improve the security in the IoT domain.

Keywords: IoT · SDN · IoT security

1 Introduction

According to the reports [1, 2] about the development of Internet of Things (IoT) domain, the growing number of things connected to the internet shows that, an emerging and large digital environment should not be underestimated. As a new generation of internet, the IoT has its special characteristics and requirements that we need to understand [3, 4].

The huge number and the heterogeneity of devices make the network devices more dynamic and complex and its management is no longer evident. Therefore, the use of a new mechanism becomes a necessity. Moreover, as we know from studies [5], the detection of abnormal behaviors is not easy in the context of network-based devices.

The traditional networks are not able to perform the automatic configuration of their components in particular the heterogeneous networks such as the network devices.

The Software Defined Networking (SDN) is a new paradigm that can facilitate and improve the management of the network devices in terms of configuration, control and security.

SDN is primarily based on the redesign of the traditional network architecture, and this via the separation of control, management and data planes. In this context, this new paradigm constitutes an opportunity to improve the IoT network security.

To address the constraints related to the network complexity in IoT domain and guaranty an efficient management of security issues, we propose, through this paper,

M. Belkasmi et al. (Eds.): ACOSIS 2019, CCIS 1264, pp. 233–238, 2020.
https://doi.org/10.1007/978-3-030-61143-9_19

the utilization of the SDN technology while fully exploiting its key features, including (i) programmability of the network; (ii) separation of data plane and control plane; (iii) a centralized controller and view of the network.

This paper is structured as follow: Sect. 2 describes the SDN paradigm. Section 3 discusses some security challenges relates to the IoT domain. Section 4 introduces how the SDN can improve the IoT security and the last Section concludes the paper.

2   Understanding the SDN Paradigm

In the traditional architecture, the network is divided into three basics components namely the management plane, the control plane and data plane [5, 6]. For the traditional network appliance, the control and the data plane are located in the same device, which makes their change and improvement difficult, if not impossible, particularly where network manufacturer of the device did not consider this aspect. However, Software Defined Networking (SDN) is a new paradigm that offers a better way to configure and manage the network through a controller, which is a central software program that controls the overall network behavior [7] and gives more information about the entire network [8]. SDN introduces a new mechanism of configuring and controlling network based on the separation of the control plane and data plane as shown in Fig. 1 proposed in [9].

Work has been undertaken to define the components of the SDN architecture, such as the work published by Open Networking Foundation (ONF) [8]. Simply, the SDN architecture is characterized by the elements above:

i. **Data plane** is composed of one or more network equipment, which contains elements of traffic forwarding or traffic processing resources.
ii. **Control plane** is the important element in this architecture, because it ensures the network (based on policies) configuration and management all of the forwarding devices.

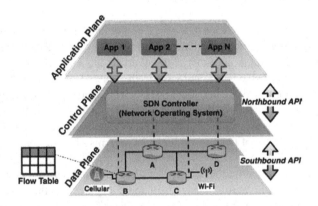

Fig. 1. SDN layers and architecture.

iii. **Application plane** is composed of one or more network applications that interact with controller and utilize abstract view of the network for their internal decision making processes.

In addition to these three layers, the SDN technology components is also based on two key elements, namely Southbound interface and Northbound interface:

iv. **Southbound interface** placed between control plane and data plane has a main role, which is to ensure the dynamic configuration of network in an easy and quick manner. OpenFlow API is a model of standardized interface in SDN architecture [10].

v. **Northbound interface** situated between control plane and application plane is responsible for providing well-defined information about routing, policy and management from controllers for application plane.

3   Insights into Security Challenges of IoT

The Internet of Things (IoT) system provides a common and large environment where the huge number of heterogeneous things will be able to communicate and interact with each other using different communication platforms. Thus, we can assess the main IoT challenges in five elements namely [11]:

- Architecture Challenge;
- Technical Challenge;
- Hardware Challenge;
- Security Challenge;
- Standard Challenge.

In this section, we will focus on the security challenges related to the network devices.

As we know, the security is the main issue from the diversity of the things and the communication of the huge number of them, which require interoperability that allows them to cooperate effectively, especially in terms of security. However, existing network security mechanisms (IDS, firewall at the gateways, IPS ...) are mainly based on static perimeter defense [12], while the network devices is characterized with dynamic behaviors that depend on variation of IoT environments and a set of device dependencies. Moreover, the IoT system needs to rapidly configure and apply new policies.

In addition, devices involved in IoT system are characterized with low processing capabilities, limited memory and storage (e.g. on flash or RAM) and minimal network protocol support [13], thus the deployment of standard security mechanisms is difficult, and it requires significant processing and storage capacities. With this feature, designing suitable security methods is a real need for more effective security.

The diversity and heterogeneity of connected devices make the traditional security systems implemented less effective or even incapable to detect and identify attacks. In addition, things that are not under a surveillance of users (Machine-to-Machine) either permanently or in a timely manner cannot identify all the sources of the attacks and their nature [11]. Thus, identification of traditional attacks such as man-in-the-middle attack, database reading attack, password cracking, etc., can be identified but the new attacks that result from the specificity of IoT paradigm, including device dependencies need new security mechanisms.

Clearly is not easy to identify all challenges related to IoT security features, but based on ideas discussed above provide insights into security challenges of IoT which show that the implementation of new mechanisms and methods in order to response to specific security requirements. In this context, the next section will hence this address how to improve IoT security through SDN technology.

4 Improving IoT Security Through SDN Technology

As discussed earlier, the specificity of Internet of Things (IoT) domain in terms of constraints, heterogeneity and interconnection of things needs to take into account the implementation of new security mechanisms in order to tackle the new challenges. In keeping with this idea, Software Defined Networking (SDN) technology is now being applied to cloud computing and virtualization domains [14]. Thus, also it can be exploited in the IoT domain, particularly to improve the IoT security mechanisms.

In the traditional network, the dynamic behaviors generated by devices are not easy to manage in the absence of monitoring approach. However, the SDN architecture can deal with this issue through the functionalities of control plane based on centralized control of complete network [8, 15]. Furthermore, the traffic and related data are analyzed in a central manner, which provides an opportunity to timely detect network wide attacks (DDoS, Spoofing, Scanning, Man-In-The-Middle ...). In this way, the implementation of security approaches, particularly the detection and analyze functions at control level will provide many benefits in terms of (i) defining adequate rules and policies without recourse to low-level IoT network (IoT devices layer); (ii) better managing of the different security requirements of various devices and applications through providing common methods and predefined rules. For example, managing the application of software patches in a centralized manner [16] via OpenFlow API.

In the same way, based on analysis of device and application behaviors in a centralized way, which is provided by some control plane functions, the SDN technology offers the opportunity to handle the trustworthiness side [16]. Thus, using an annotation system, controllers can easily define the trust level of entities involved in IoT system [17]. Note that the trust assessment is a key element of efficient management of device dependencies.

The availability of IoT devices and services is an elementary requirement for quality of service in IoT paradigm, in this vein, the decoupling of the control and forwarding functions can effectively address the availability requirement. Furthermore, decoupling security software from hardware is an opportunity to exceed the challenge related to the recourse constraints of IoT devices.

5 Conclusion and Future Work

The aspect of security in Internet of Things (IoT) is both important and complex, in this context the main characteristics of IoT system that consist of diversity, unlimited network of devices and resource constraints need to be taken in consideration. In this paper, we tried to throw light on some security challenges that do not allow things to act in a more efficient and autonomous way. On the other hand, we have discussed the added value of the Software Defined Networking (SDN) paradigm to improve the IoT security. Moreover, unlike the traditional network, the new concept of SDN based on decoupling control and data planes can address some security requirements and ensure the quality of service and security. Regarding the benefits discussed above they primarily concern the monitoring, management and of network devices in a manner the decoupling of control and forwarding functions. For the future work, we will address the implementation of these approaches in IoT domain, including the design of SDN-based IoT architecture while fully exploiting its benefits in terms of centralized management and dynamic reconfiguration.

References

1. Nag-Chowdhury, S., Manohar-Kuhikar, K., Dhawan, S.: IoT architecture: a survey. In: 21st IRF International Conference, pp. 35–38 (2015)
2. Vermesan, O., et al.: Internet of Things strategic research and innovation agenda. In: Vermesan, O., Friess, P. (eds.) Internet of Things - Converging Technologies for Smart Environments and Integrated Ecosystems, pp. 7–152. River Publishers (2013)
3. Ghasempour, A.: Internet of Things in smart grid: architecture, applications, services, key technologies, and challenges. Invent. J. **4**(1), 1–12 (2019)
4. Ghasempour, A.: Optimum number of aggregators based on power consumption, cost, and network lifetime in advanced metering infrastructure architecture for smart grid Internet of Things. In: IEEE Annual Consumer Communications & Networking Conference, Las Vegas, pp. 295-6 (2016)
5. Ferrando, R., Stacey, P.: Classification of device behaviour in Internet of Things infrastructures: towards distinguishing the abnormal from security threats. In: International Conference on Internet of Things and Machine Learning, Liverpool, pp. 1–8 (2017)
6. Durand, J.: Le SDN pour les nuls. 11e Journées RESeaux (JRES), pp. 1–12 (2015)
7. Kim, H., Feamster, N.: Improving network management with software defined networking. In: IEEE Communications Magazine, pp. 14–119 (2013)
8. Open Networking Foundation (eds.): SDN Architecture, pp 1–68 (2014)
9. Shaghaghi, A., Ali-Kaafar, M., Buyya, R., Jha, S.: Software-defined network (SDN) data plane security: issues, solutions and future directions. J. Cluster Comput. pp. 1–24 (2018)
10. Shin, M., Nam, K., Kim, H.: The software defined networking (SDN): a reference architecture and open APIs. In: 2012 International Conference on ICT Convergence (ICTC), pp. 360–361 (2012)
11. Patel, K.K., Patel, S.M.: Internet of Things-IOT: definition, characteristics, architecture, enabling technologies, application & future challenges. Int. J. Eng. Sci. Comput. **6**, 6122–6131 (2016)

12. Yu, T., Sekar, V., Seshan, S., Agarwal, Y., Xu, C.: Handling a trillion (unfixable) flaws on a billion devices: Rethinking network security for the Internet-of-Things. In: HotNets 2015, pp. 1–7 (2015)
13. Mocana Blog: 5 key challenges in securing resource-constrained IoT devices. www.mocana.com/blog/5-key-challenges-in-securing-resource-constrained-iot-devices
14. Tsugawa, M., Matsunaga, A., Fortes, J.: Cloud computing security: what changes with software-defined networking? Secure Cloud Comput. **2014**, 77–93 (2014)
15. Hu, Z., Wang, M., Yan, X., Yin, Y., Luo, Z.: A comprehensive security architecture for SDN. In: 18th International Conference on Intelligence in Next Generation Networks, pp. 30–37 (2015)
16. Ferraris, D., Fernandez-Gago, C., Lopez, J.: A trust-by-design framework for the Internet of Things. In: 9th IFIP International Conference on New Technologies, Mobility and Security (NTMS), pp. 1–4 (2018)
17. Sicari, S., Cappiello, C., De Pellegrini, F., Miorandi, D., Coen-Porisini, A.: A security-and quality-aware system architecture for Internet of Things. Inf. Syst. Front. **18**(4), 665–677 (2014). https://doi.org/10.1007/s10796-014-9538-x

Applied Cryptography

Optical Image Encryption Process Using Triple Deterministic Spherical Phase Masks Array

Wiam Zamrani[(✉)] and Esmail Ahouzi

Institut National des Postes et Télécommunications, Rabat 10000, Morocco
zamrani.wiam@gmail.com

Abstract. In this paper, we perform a study of optical image encryption using triple deterministic phase mask (TDPE) based on diffractive spherical phase masks array (SPMA). By combining the diffractive spherical phase masks array with the deterministic phase masks method we can improve the security level of an encryption scheme by increasing the key space and providing additional security parameters and also by overcoming the problem of the sensitivity to misalignment associated with an optical setup which is beneficial for practical implementation. Numerical simulation results indicate that the proposed method can reach a better effect in image encryption compared with the conventional double random phase encryption (DRPE).

Keywords: Optical security · Encryption · Deterministic phase mask

1 Introduction

The security of data (audio, video, image, etc.) has received increasingly more attention in recent years owing to the continuous evolution of technologies. Many optical encryption techniques have been developed recently because of their property of parallel processing of a large amount of data at high speed [1–6]. In 1995, Refregier and Javidi [7] presented the double random phase encryption scheme (DRPE) which is considered as the first pioneering work in optical image security. The DRPE consists of encoding the input original image into white noise image employing two generated random phase masks (RPMs) under the classical 4f correlator. Practical disadvantages of DRPE occur when attempting to position the (RPMs) in decryption phase; the original information cannot be recovered due to the optical mispositioning. For the purpose to enhance the security level of DRPE, it was draw out to different canonical transform domains, for instance, fractional Fourier transform [8], Hartley transform [9], Fresnel [10], gyrator [11], and fraction Mellin [12]. However, these encryption methods are susceptible to some special attacks [13–16] because of the linear characteristic of their transformations. To decrease the potential risk occurring from the linear transform algorithms, several encoding schemes based on nonlinear processing have been proposed [17–21].

As an alternative to the RPMs used on the conventional DRPE scheme, several researchers proposed to use a different kind of phase mask named structured phasemasks (SPMs) generated based on different techniques like the Fresnel zone-plate (FZP), the toroidal zone-plate (TZP), and the radial-hilbert mask (RHM). The use of

© Springer Nature Switzerland AG 2020
M. Belkasmi et al. (Eds.): ACOSIS 2019, CCIS 1264, pp. 241–250, 2020.
https://doi.org/10.1007/978-3-030-61143-9_20

SPMs was first presented by Barrera in 2005 [22, 23] who proved that by using SPMs such as the toroidal zone plate (TZP) will help in the exact positioning in the decryption process. In 2009, Tebaldi et al. [24] proposed the use of diffractive optical elements, for instance the fractal zone plate (FrZP) which expanded the robustness of the encryption system and ensured a high degree of versatility. In 2013, Aburturab [25] introduced an information security scheme for colored images, using Arnold transform and double SPMs in the gyrator domain, for the purpose to increase the key size and thus, improve the security level of the cryptosystem, the structured phase masks are generated based on the phase function of FZP. In 2014, Vashisth et al. [26] proposed an encryption scheme for grayscale images based on the phase retrieval technique in the fractional Mellin domain, and two SPMs that are generated based on TZP for the first key, and RHM for the second key, as a result, the method enlarge the key space and increase the security level of the scheme. In addition to different SPMs cited above, several researchers proposed the use of another type of SPMs called devil's vortex Fresnel lens (DVFL) that also improves the security of an encryption algorithm by adding extra-parameters to the generation of the masks [27–29].

In this paper, we tried to merge our recently proposed image encryption techniques in Ref. [30, 31] by using a new deterministic configuration for the encryption—decryption keys, which lead to the triple deterministic spherical phase masks array encryption process. This process uses triple SPMs in order to enhance the security level against different attacks, and these SPMs are generated based on a special configuration that relies on specific parameters. Numerical simulations have been carried out to test the effectiveness and credibility of our proposed technique. The results proved that our proposed technique crucially improves the misalignment that occurs on the conventional DRPE.

The paper is organized as follows. Section 2 describes the principle of encryption and decryption of our proposed cryptosystem. The viability and performance analysis of our proposed encryption technique are verified by numerical simulations in Sect. 3. Finally, Sect. 4 presents the conclusions of the work.

2   System Principles

In this section, we propose to implement the optical triple phase mask cryptosystem in Ref. [31] by using the deterministic spherical phase masks described later in this section. The operations of encryption and decryption process are illustrated in Fig. 1 and Fig. 2, respectively.

Allow that $f(x,y)$ stand for the input image to be encrypted and, SPM1, SPM2 and SPM3 the three deterministic spherical phase masks used in the three plane of the VanderLugt correlator, respectively, as indicated in Fig. 3. In the encryption process, the input original image $f(x,y)$ is multiplied with the first deterministic spherical phase-mask $SPM_1(x,y)$, this product is Fourier transformed by lens 1, then multiplied with the second deterministic spherical phase mask $SPM_2(u,v)$ at the Fourier plane, another Fourier transformation is performed by lens 2, and the result is multiplied by the third deterministic spherical phase mask $SPM_3(x,y)$, which lead to the output encrypted image $e(x,y)$. The encryption process described above is described by:

$$e\,(x,y) = \{\text{IFT}\{\text{IFT}\{f(x,y) \times SPM_1(x,y)\} \times SPM_2(u,v)\}\} \times SPM_3(x,y) \quad (1)$$

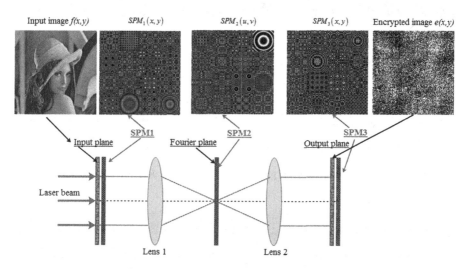

Fig. 1. Optical scheme for encryption process.

To decrypt $e(x,y)$ the reverse process of encryption is applied as presented in Fig. 2, where the encoded image $e(x,y)$ is multiplied by the conjugation of the third deterministic spherical phase mask $SPM_3{}^*$ (the symbol * means the complex conjugate). Then, the inverse Fourier transform is performed by lens 1. Thus, the result is then multiplied by the conjugation of the second deterministic spherical phase mask $SPM_2{}^*$. Then, lens2 performs another inverse Fourier transform on the obtained image, this transformation lead to the complex product of the input image $f(x,y)$ with the first mask used in the encryption process $SPM_1{}^*$. The decryption process can be described by:

$$f'(x,y) = \text{IFT}\{\text{IFT}[e(x,y) \times SPM_3^*(x,y)] \times SPM_2^*(u,v)\} \times SPM_1^*(x,y) \quad (2)$$

The deterministic spherical phase masks are obtained as follow:

$$\text{SPM} = \sum_{i=1}^{2m} \sum_{j=1}^{2m} M_{ij}(x,y) \quad (3)$$

$$M_{ij}(x,y) = \exp\{-i2\pi[u_k(x^2)/2 + v_k(y^2)/2]\}, k = 1, \ldots, (2^m \times 2^m) \quad (4)$$

Where m defines the order of encryption. In our case we will consider m = 2.

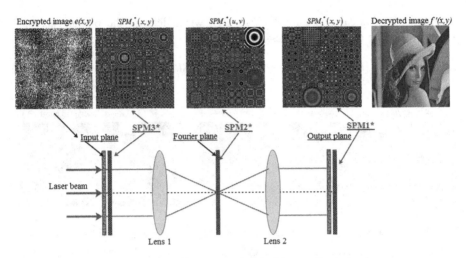

Fig. 2. Schematic process of optical image decryption.

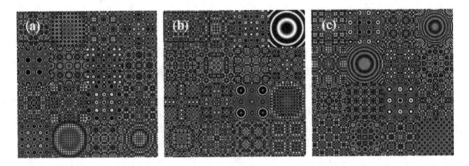

Fig. 3. Spherical phase masks used in the encryption/decryption process.

3 Numerical Simulation and Performance Analysis

To study the suitability, robustness and sensitivity of our proposed scheme, numerical simulations were carried out on the Matlab2014a platform on the images with different formats (JPG and PNG) of size 256 × 256 pixels. The images we have used for testing our algorithm are shown in Figs. 4(a) and 4(b). In addition, Figs. 4(c) and 4(d) display the encrypted images that correspond to the original images respectively. The decrypted images are shown in Figs. 4(e) and 4(f). From the results shown in Fig. 4, all the encrypted images are completely unrecognizable. It shows that the proposed encryption method has a good encryption effect.

Fig. 4. Simulation results of the scheme: (a) and (b) original images, (c) and (d) their encrypted images, (e) and (f) the corresponding decrypted images.

To evaluate the image quality, the difference between the original image and the decrypted one is measured by calculating the Peak signal to noise (PSNR) and the Mean square error (MSE) metrics, expressed by:

$$
\text{MSE} = \frac{1}{N \times M} \sum_{x=0}^{N-1} \sum_{y=0}^{M-1} |f'(x,y) - f(x,y)|^2 \tag{5}
$$

$$
\text{PSNR} = 10 \log_{10} \left(\frac{255^2}{\sqrt{MSE}} \right) \tag{6}
$$

In this case, $f'(x,y)$ represent the recovered image and $f(x,y)$ the original image. The computed values of MSE of the input and the recovered images using the proposed scheme for "Lena" image and "binary text" image are 1.23×10^{-31} and 2.29×10^{-32} respectively. The PSNR values between input and decrypted images for Lena and binary text are 357.12, 364.51 dB respectively. The result values are listed in Table 1 in comparison with the conventional DRPE encryption scheme.

Table 1. The results of computing the MSE and PSNR metrics.

Encryption algorithms	MSE		PSNR (dB)	
	Lena	Binary text	Lena	Binary text
Conventional DRPE [7]	1.23×10^{-31}	2.33×10^{-32}	357.18	364.45
Our proposed method	1.23×10^{-31}	2.29×10^{-32}	357.12	364.51

Table 1 shows the results values obtained by computing the MSE and PSNR between the original and the retrieved images. As can be seen, the values are similar between our proposed method and the standard DRPE. Consequently, we can say that both encryption techniques have the same degree of security and are equivalent when using the exact encryption-decryption keys.

3.1 Robustness to Data Loss and Noise Attacks

The robustness against data loss on the encrypted images is examined. The occluded images are shown in Figs. 5(a)–(d) and 5(i)–(l) for 65×65, 85×85, 128×128 and 200×200 pixels size occlusion in encrypted images of Lena and binary text, respectively. Figures 5(e)–(h) and 5(m)–(p) represent the related recovered images that are well retrieved even for occlusion higher than 200×200 pixels size. Obviously, the decrypted images still contain most of the original visual information. So the proposed encryption method is robust against the data loss. The calculated values of MSE metric between the original images and their corresponding recovered images are recorded in Table 2.

Table 2. The MSE values between original and retrieved images to data loss.

Input images	Data loss	MSE	
		DRPE	Our method
Lena image	65×65	0.0186	0.0258
	85×85	0.0309	0.0333
	128×128	0.0717	0.0519
	200×200	0.1724	0.1294
Text image	65×65	0.0035	0.0027
	85×85	0.0061	0.0033
	128×128	0.0135	0.0047
	200×200	0.0330	0.0204

When compared to the numerical simulation results, it is obvious that our proposed method has almost the same behaviors as the conventional DRPE, and can reconstruct a whole scene of images. As can be seen, from the simulation tests of Fig. 5, it is obvious that our proposed scheme shows higher robustness compared to DRPE. The calculation of MSE values are tabulated in Table 2. The MSE values of the proposed method are improved compared for those of DRPE algorithms under the occlusion attack.

We investigate also the tolerance of our proposed scheme to the noise attacks. For noise attacks, Gaussian noise, Speckle noise and Salt & Pepper noise are added into the encrypted images of "Lena" and "binary text". Making use of MSE metric to test the quality of the recovered images, the simulation results of the three kinds of noise attacks are recorded in Table 3. When compared to the conventional DRPE scheme, the simulation results shows that the MSE values of the proposed method are the same as DRPE which indicate that our proposed scheme is sensitive to noise attacks.

Fig. 5. Data loss results for the proposed scheme: Encrypted images of Lena (a)–(d) and binary text (i)–(l) with different pixels sizes of information loss: (a) and (i) 65 × 65 pixels, (b) and (j) 85 × 85 pixels, (c) and (k) 128 × 128 pixels and (d) and (l) 200 × 200 pixels. Their corresponding retrieved images of Lena in (e, f, g, h) and binary image in (m, n, o, p).

Table 3. The MSE values between original and retrieved images under Noise Attacks.

Input images	Noise Attacks	MSE	
		DRPE	Our method
Lena image	Gaussian noise	0.2835	0.2835
	Speckle noise	0.2835	0.2835
	Salt & Pepper	0.2835	0.2835
Text image	Gaussian noise	0.0543	0.0543
	Speckle noise	0.0543	0.0543
	Salt & Pepper	0.0543	0.0543

3.2   Robustness to Spatial Shift

During this section, we evaluate the effectiveness of our proposed scheme to shift tolerance then we will compare the results with the standard DRPE. Figure 6 represents the recovered images obtained with our encryption technique when the second phase mask $SPM_2*(u,v)$ in the Fourier plane is shifted from its position through x and y directions in steps of one (Fig. 6(a)), two (Fig. 6(b)) and three (Fig. 6(c)) pixels for the "Lena" image. Same results are respectively provided in Figs. 6(d), 6(e), and 6(f) for the "binary text" image. Consequently, even when the retrieved phase mask is shifted from the matching position along the x-y axis, the recovered images (Fig. 6) are still recognizable.

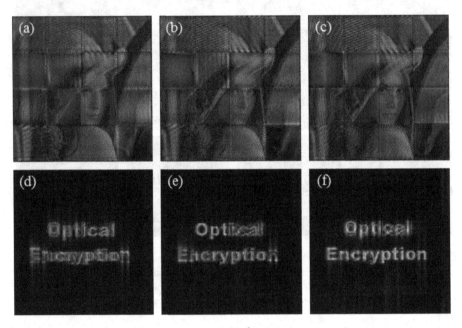

Fig. 6. Decrypted images obtained when the SPM_2^* is shifted from the original position by one (a) and (d), two (b) and (e), and three pixels (c) and (f) for Lena and binary text image.

Table 4 displays the MSE values of the original images end the corresponding decrypted images by our proposed encryption method and the conventional DRPE. The quantitative results show that our proposed encryption scheme has a small sensitivity to pixels shift in comparison with the DRPE.

Table 4. The MSE results between original and retrieved images under pixels shift.

Input images	Pixels shift	MSE	
		DRPE	Our method
Lena image	1 pixel	0.5667	0.0248
	2 pixels	0.5648	0.0277
	3 pixels	0.5645	0.0280
Text image	1 pixel	0.1081	0.0259
	2 pixels	0.1086	0.0232
	3 pixels	0.1087	0.0216

4   Conclusion

In this paper, an optical image encryption technique is presented namely, the triple deterministic spherical phase masks array. The parameters in spherical phase mask array (SPMA) serve as additional keys to increase the security of our proposed encryption process. The use of SPMA alleviates the alignment problem and provides high security level. To test the feasibility of our proposed scheme, we computed the Mean square error and the peak signal to noise between the original images and their corresponding recovered images. Numerical simulation results demonstrate the validity, security, and the strength of the proposed cryptosystem.

References

1. Javidi, B.: Optical and Digital Techniques for Information Security, pp. 241–269. Springer, New York (2005). https://doi.org/10.1007/b105098
2. Chen, W., Javidi, B., Chen, X.: Advances in optical security systems. Adv. Opt. Photon. **6**, 120–155 (2014)
3. Matoba, O., Nomura, T., Pérez-Cabré, E., Millán, M.S., Javidi, B.: Optical techniques for information security. Proc. IEEE **97**, 1128–1148 (2009)
4. Li, Y., Kreske, K., Rosen, J.: Security and encryption optical systems based on a correlator with significant output images. Appl. Opt. **39**(29), 5295–5301 (2000)
5. Liu, S., Guo, C., Sheridan, J.T.: A review of optical image encryption techniques. Opt. Laser Technol. **57**, 327–342 (2014)
6. Javidi, B., et al.: Roadmap to optical security. J. Opt **18**, 083001 (2016)
7. Refregier, P., Javidi, B.: Optical image encryption based on input plane and Fourier plane random encoding. Opt. Lett. **20**(7), 767–769 (1995)
8. Unnikrishnan, G., Joseph, J., Singh, K.: Optical encryption by double-random phase encoding in the fractional Fourier domain. Opt. Lett. **25**, 887–889 (2000)
9. Chen, L., Zhao, D.: Optical image encryption with Hartley transforms. Opt. Lett. **31**, 3438–3440 (2006)
10. Situ, G., Zhang, J.: Double random-phase encoding in the Fresnel domain. Opt. Lett. **29**, 1584–1586 (2004)
11. Singh, H., Yadav, A.K., Vashisth, S., Singh, K.: Fully phase image encryption using double random structured phase masks in gyrator domain. Appl. Opt. **53**, 6472–6481 (2014)

12. Zhou, N.R., Wang, Y., Gong, L.: Novel optical image encryption scheme based on fractional Mellin transform. Opt. Commun. **284**, 3234–3242 (2011)
13. Peng, X., Zhang, P., Wei, H., Yu, B.: Known-plaintext attack on optical encryption based on double random phase keys. Opt. Lett. **31**, 1044–1046 (2006)
14. Peng, X., Wei, H., Zhang, P.: Chosen-plaintext attack on lensless double-random phase encoding in the Fresnel domain. Opt. Lett. **31**, 3261–3263 (2006)
15. Carnicer, A., et al.: Vulnerability to chosen-ciphertext attacks of optical encryption schemes based on double random phase keys. Opt. Lett. **30**(13), 1644–1646 (2005)
16. Unnikrishnan, G., et al.: A known-plaintext heuristic attack on the Fourier plane encryption algorithm. Opt. Express **14**(8), 3181–3186 (2006)
17. Vilardy, O.J.M., Perez, R.A., Torres, M.C.O.: Optical image encryption using a nonlinear joint transform correlator and the collins diffraction transform. Photonics **6**, 115
18. Shan, T., Chen, T., Shen, Y., Zhenkun, L.: Optical image encryption based on biometric keys and singular value decomposition. Appl. Opt. **59**, 2422–2430 (2020)
19. Chen, J., Bao, N., Zhang, L., Zhu, Z.: Optical information authentication using optical encryption and sparsity constraint. Opt. Lasers Eng. **107**, 352–363 (2018)
20. Jiao, S., Zhuang, Z., Zhou, C., Zou, W., Li, X.: Security enhancement of double random phase encryption with a hidden key against ciphertext only attack. Opt. Commun. **418**, 106–114 (2018)
21. Chen, J., Zhang, Y., Li, J., Zhang, L.B.: Security enhancement of double random phase encoding using rear-mounted phase masking. Opt. Laser. Eng. **101**, 51–59 (2018)
22. Barrera, J.F., Henao, R., Torroba, R.: Optical encryption method using toroidal zone plates. Opt. Commun. **248**, 35–40 (2005)
23. Barrera, J.F., Henao, R., Torroba, R.: Fault tolerances using toroidal zone plate encryption. Opt. Commun. **256**, 489–494 (2005)
24. Tebaldi, M., Furlan, W.D., Torroba, R., Bolognini, N.: Optical-data storage-readout technique based on fractal encrypting masks. Opt. Lett. **34**, 316–318 (2009)
25. Aburturab, M.R.: Color information security system using Arnold transform and double structured phase encoding in gyrator transform domain. Opt. Laser Technol. **45**, 524–532 (2013)
26. Vashisth, S., Singh, H., Yadav, A.K., Singh, K.: Image encryption using fractional Mellin transform, structured phase filters, and phase retrieval. Optik **125**, 5309–5315 (2014)
27. Mitry, M., Doughty, D.C., Chaloupka, J.L., Anderson, M.E.: Experimental realization of the devil's vortex Fresnel lens with a programmable spatial light modulator. Appl. Opt. **51**, 4103–4108 (2012)
28. Calatayud, A., Rodrigo, J.A., Remon, L., Furlan, W.D., Cristobal, G., Monsoriu, J.A.: Experimental generation and characterization of Devil's vortex-lenses. Appl. Phys. B **106**, 915–919 (2012)
29. Calabuig, A., et al.: Generation of programmable 3D optical vortex structures through devil's vortex-lens arrays. Appl. Opt. **52**, 5822–5829 (2013)
30. Zamrani, W., Ahouzi, E., Lizana, A., Campos, J., Yzuel, J.M.: Optical image encryption technique based on deterministic phase masks. Opt. Eng. **55**(10), 103108 (2016)
31. Ahouzi, E., Zamrani, W., Azami, N., Lizana, A., Campos, J., Yzuel, J.M.: Optical triple random-phase encryption. Opt. Eng. **56**(11), 113114 (2017)

Author Index

Abdelmounim, Elhassane   115
Abdulrahman, Noora   53
Addaim, Adnane   180
Ahouzi, Esmail   241
Alnajjar, Khawla A.   53
Alramsi, Marwah   53
Amar, Meryem   170
Ar-Reyouchi, El Miloud   38
Ayoub, Fouad   121
Azougaghe, Es-said   103

Belhabib, Abdelfettah   3
Belkasmi, Mostafa   95, 103, 121
Benaddi, Hafsa   205
Benchaib, Imane   38
Benjbara, Chaimae   219
Benslimane, Abderrahim   205
Boualame, Hamza   95
Boudaoud, Abdelghani   115
Bouderba, Saif Islam   65
Boulouird, Mohamed   3
Boumaouche, Oualid   195
Bousarhane, Btissam   73
Bouzidi, Driss   73

Chana, Idriss   95

Dahri, Haytham   205
Damej, Loubna   180
Diouri, Ouafaa   233

El Haroussi, Mustapha   115
EL Ouahidi, Bouabid   170

Farchane, Abderrazak   103

Ghenai, Afifa   195
Ghoumid, Kamal   38

Habbani, Ahmed   219
Hassani, Moha M'Rabet   3

Ibrahimi, Khalil   205

Karbal, Basma   157
Kerrich, Mohamed Amine   180
Kirmizioglu, Riza Arda   25

Laaouine, Jamal   137
Lamrani, Yousra   38

M'rabet, Zakaria   121
Mahdi, Fatma   53
Merah, Hocine   14
Mesri, Mokhtaria   14
Mouchfiq, Nada   219
Moussa, Najem   65

Rattal, Salma   38
Romadi, Rahal   157

Safi, Said   103

Tahkoubit, Khaled   14
Talbi, Larbi   14
Tekalp, A. Murat   25
Tioutiou, Abdelaali   233

Zamrani, Wiam   241
Zeghib, Nadia   195

Printed in the United States
By Bookmasters